Advances in Communications-Based Train Control Systems

OTHER COMMUNICATIONS BOOKS FROM CRC PRESS

Analytical Evaluation of Nonlinear Distortion Effects on Multicarrier Signals
Theresa Araújo
ISBN 978-1-4822-1594-6

Architecting Software Intensive Systems: A Practitioners Guide
Anthony J. Lattanze
ISBN 978-1-4200-4569-7

Cognitive Radio Networks: Efficient Resource Allocation in Cooperative Sensing, Cellular Communications, High-Speed Vehicles, and Smart Grid
Tao Jiang, Zhiqiang Wang, and Yang Cao
ISBN 978-1-4987-2113-4

Complex Networks: An Algorithmic Perspective
Kayhan Erciyes
ISBN 978-1-4665-7166-2

Data Privacy for the Smart Grid
Rebecca Herold and Christine Hertzog
ISBN 978-1-4665-7337-6

Generic and Energy-Efficient Context-Aware Mobile Sensing
Ozgur Yurur and Chi Harold Liu
ISBN 978-1-4987-0010-8

Just Ordinary Robots: Automation from Love to War
Lamber Royakkers and Rinie van Est
ISBN 978-1-4822-6014-4

Machine-to-Machine Communications: Architectures, Technology, Standards, and Applications
Vojislav B. Misic and Jelena Misic
ISBN 978-1-4665-6123-6

Managing the PSTN Transformation: A Blueprint for a Successful Migration to IP-Based Networks
Sandra Dornheim
ISBN 978-1-4987-0103-7

MIMO Processing for 4G and Beyond: Fundamentals and Evolution
Edited by Mário Marques da Silva and Francisco A. Monteiro
ISBN 978-1-4665-9807-2

Mobile Evolution: Insights on Connectivity and Service
Sebastian Thalanany
ISBN 978-1-4822-2480-1

Network Innovation through OpenFlow and SDN: Principles and Design
Edited by Fei Hu
ISBN 978-1-4665-7209-6

Neural Networks for Applied Sciences and Engineering: From Fundamentals to Complex Pattern Recognition
Sandhya Samarasinghe
ISBN 978-0-8493-3375-0

Rare Earth Materials: Properties and Applications
A.R. Jha
ISBN 978-1-4665-6402-2

Requirements Engineering for Software and Systems, Second Edition
Phillip A. Laplante
ISBN 978-1-4665-6081-9

Security for Multihop Wireless Networks
Edited by Shafiullah Khan and Jaime Lloret Mauri
ISBN 978-1-4665-7803-6

The Future of Wireless Networks: Architectures, Protocols, and Services
Edited by Mohesen Guizani, Hsiao-Hwa Chen, and Chonggang Wang
ISBN 978-1-4822-2094-0

The Internet of Things in the Cloud: A Middleware Perspective
Honbo Zhou
ISBN 978-1-4398-9299-2

The State of the Art in Intrusion Prevention and Detection
Al-Sakib Khan Pathan
ISBN 978-1-4822-0351-6

ZigBee® Network Protocols and Applications
Edited by Chonggang Wang, Tao Jiang, and Qian Zhang
ISBN 978-1-4398-1601-1

TO ORDER
Call: 1-800-272-7737 • Fax: 1-800-374-3401 • E-mail: orders@crcpress.com

Advances in Communications-Based Train Control Systems

Edited by F. Richard Yu

CRC Press
Taylor & Francis Group
Boca Raton London New York

CRC Press is an imprint of the
Taylor & Francis Group, an **informa** business

CRC Press
Taylor & Francis Group
6000 Broken Sound Parkway NW, Suite 300
Boca Raton, FL 33487-2742

First issued in paperback 2018

© 2016 by Taylor & Francis Group, LLC
CRC Press is an imprint of Taylor & Francis Group, an Informa business

No claim to original U.S. Government works

ISBN-13: 978-1-4822-5743-4 (hbk)
ISBN-13: 978-1-138-89450-1 (pbk)

Contents

List of Figures..vii
List of Tables xiii
Preface xv
Editor..xix
Contributors...xxi

1 Introduction to Communications-Based Train Control 1
LI ZHU, F. RICHARD YU, AND FEI WANG

2 Testing Communications-Based Train Control 15
KENNETH DIEMUNSCH

3 Safe Rail Transport via Nondestructive Testing Inspection
of Rails and Communications-Based Train Control Systems 43
VASSILIOS KAPPATOS, TAT-HEAN GAN, AND DIMITRIS
STAMATELOS

4 Modeling of the Wireless Channels in Underground Tunnels
for Communications-Based Train Control Systems..............................65
HONGWEI WANG, F. RICHARD YU, LI ZHU, AND TAO TANG

5 Modeling of the Wireless Channels with Leaky Waveguide
for Communications-Based Train Control Systems..............................81
HONGWEI WANG, F. RICHARD YU, LI ZHU, AND TAO TANG

6 Communication Availability in Communications-Based Train
Control Systems...93
LI ZHU AND F. RICHARD YU

7 Novel Communications-Based Train Control System
with Coordinated Multipoint Transmission and Reception 115
LI ZHU, F. RICHARD YU, AND TAO TANG

8 Novel Handoff Scheme with Multiple-Input and Multiple-Output for Communications-Based Train Control Systems............................ 149
HAILIN JIANG, VICTOR C. M. LEUNG, CHUNHAI GAO, AND TAO TANG

9 Networked Control for a Group of Trains in Communications-Based Train Control Systems with Random Packet Drops 177
BING BU, F. RICHARD YU, AND TAO TANG

10 Cognitive Control for Communications-Based Train Control Systems 213
HONGWEI WANG AND F. RICHARD YU

Index 247

List of Figures

Figure 1.1 Train signaling system using wayside signals 2

Figure 1.2 Profile-based train control system 4

Figure 1.3 CBTC system..4

Figure 1.4 Typical architecture of a modern CBTC system...............................6

Figure 3.1 Superstructure subsection 47

Figure 3.2 Typical crack propagation in the head area of rail 48

Figure 3.3 (a) Broken rail, (b) examples of severe loss of rail foot due to severe corrosion, and (c) shelling...48

Figure 3.4 (a) Common rigid and (b) elastic fastening 50

Figure 3.5 Manual rail inspections expose maintenance personnel to dangers from passing trains, flying ballast, and projectiles 51

Figure 3.6 Coverage of train-mounted and walking stick probes (a), geometrical limitations on the current ultrasonic inspection (b), and example of a rail break due to small rail foot defect (c)...........53

Figure 3.7 Transducer mounting for rail foot inspection................................55

Figure 4.1 Measurement equipment used in the real field CBTC channel measurements 68

Figure 4.2 Tunnel section and deployment of antennas in the measurement..69

Figure 4.3 (a) Tunnel where we performed the measurements in Beijing Subway Changping Line. (b) Shark-fin antenna located on the measurement vehicle. (c) Yagi antenna. (d) AP set on the wall 70

Figure 4.4 Frequencies of AICc selecting a candidate distribution 75

Figure 4.5 Simulation results generated from the FSMC model and experimental results from real field measurements 78

Figure 4.6 MSE between the FSMC model and the experimental data with four states and eight states..78

Figure 5.1 Leaky waveguide applied in viaduct scenarios of Beijing Subway Yizhuang Line...83

Figure 5.2 Measurement equipment used in the CBTC channel measurements 84

Figure 5.3 Measurement scenario... 84

Figure 5.4 Simulation results of the equivalent method and the fitting lines of several measurements..87

Figure 5.5 Relative frequencies of AICc selecting a candidate distribution as the best fit to the distribution of small-scale fading amplitudes... 89

Figure 5.6 Sample empirical CDFs of the small-scale fading amplitudes and their theoretical model fits 89

Figure 5.7 Variance of μ_{dB} and σ_{dB} with different receiving points................. 90

Figure 6.1 CBTC system 96

Figure 6.2 First proposed data communication system with redundancy and no backup link..98

Figure 6.3 Second proposed data communication system with redundancy and backup link..99

Figure 6.4 CTMC model for the data communication system with basic configuration 100

Figure 6.5 CTMC model for the first proposed data communication system with redundancy and no backup link 100

Figure 6.6 CTMC model for the first proposed data communication system with redundancy and backup link 102

Figure 6.7 DSPN model for the data communication system with basic configuration 104

Figure 6.8 DSPN model for the data communication system with redundancy and no backup link..105

Figure 6.9 DSPN model for the proposed data communication system with redundancy and backup link...105

Figure 6.10   Comparison of CTMC and DSPN model solutions for different redundancy configurations 111

Figure 6.11   Unavailability of the three WLAN-based data communication systems 112

Figure 7.1   Impacts of wireless communications on CBTC efficiency 120

Figure 7.2   Trip error under different handoff communication latencies 121

Figure 7.3   Proposed CBTC system with CoMP 122

Figure 7.4   Train control model 123

Figure 7.5   Optimal train guidance trajectory 137

Figure 7.6   Control performance H_2 norm in different schemes 141

Figure 7.7   Train travel trajectory in the proposed CBTC system with CoMP 142

Figure 7.8   Train travel trajectory in the existing CBTC system 142

Figure 7.9   Train travel time error in different schemes 143

Figure 7.10   Handoff policies in different schemes 144

Figure 7.11   Average service discontinuity time duration in different schemes 145

Figure 8.1   WLAN handoff timing diagram 152

Figure 8.2   Proposed handoff scheme 156

Figure 8.3   FER of different transmission schemes 159

Figure 8.4   Delay difference between adjacent APs 160

Figure 8.5   Latency performance improvements of the MAHO scheme 168

Figure 8.6   Time interval between two handoff procedures, $v = 80$ km/h ... 169

Figure 8.7   (a) FER at 6.5 Mbps. (b) FER at 13 Mbps. (c) FER at 26 Mbps. (d) FER at 65 Mbps with different coverage areas 170

Figure 8.8   Maximum inter-site distance to meet the HO FER 173

Figure 9.1   Train following model 181

Figure 9.2   Communication procedure between ZC and the running trains with packet drops 181

Figure 9.3   Model of system to control a group of trains in CBTC 182

Figure 9.4   Equivalent networked control system...183

Figure 9.5   FER at certain train speeds 186

Figure 9.6   Probabilities that the received power at a given distance exceeds certain levels..190

Figure 9.7   Overlapping coverage area of APs..191

Figure 9.8   Field test results on handover time..192

Figure 9.9   Performances of T1 and T2 using the Sv scheme, $T = 0.3$ s, $P(\gamma_k^i = 0) = P(\theta_k^i = 0) = 0.01$..205

Figure 9.10 Performances of T3 using the Sv scheme, $T = 0.3$ s, $P(\gamma_k^i = 0) = P(\theta_k^i = 0) = 0.01$..205

Figure 9.11 Performances of T3 using the Lv_d scheme, $T = 0.3$ s, $P(\gamma_k^i = 0) = P(\theta_k^i = 0) = 0.01$.. 206

Figure 9.12 Performances of T3 using the Lv_f scheme, $T = 0.3$ s, $P(\gamma_k^i = 0) = P(\theta_k^i = 0) = 0.01$.. 206

Figure 9.13 Performances of T3 using the Sv scheme, $T = 0.3$ s, $P(\gamma_k^i = 0) = P(\theta_k^i = 0) = 0.1$ 207

Figure 9.14 Performances of T3 using the Lv_d scheme, $T = 0.3$ s, $P(\gamma_k^i = 0) = P(\theta_k^i = 0) = 0.1$ 207

Figure 9.15 Performances of T3 using the Lv_f scheme, $T = 0.3$ s, $P(\gamma_k^i = 0) = P(\theta_k^i = 0) = 0.1$ 208

Figure 9.16 Cost of T3 using the Sv, Lv_f, and Lv_d schemes, respectively, $T = 0.3$ s, $P(\gamma_k^i = 0) = P(\theta_k^i = 0) = 0.1, 0.01$ 209

Figure 9.17 Total cost of T3 using the Sv, Lv_f, and Lv_d schemes, respectively, $T = 0.3$ s, $P(\gamma_k^i = 0) = P(\theta_k^i = 0) = 0.1$ 209

Figure 10.1 Basic schematic structure of a cognitive control system 216

Figure 10.2 Basic procedure of a cognitive control system............................217

Figure 10.3 RL model...219

Figure 10.4 Schematic structure of the cognitive control approach to CBTC systems 222

Figure 10.5 RL model in the cognitive control approach 225

Figure 10.6 Basic WLAN handoff procedure.. 228

Figure 10.7   (a) Tunnel where we performed the measurements.
(b) Shark-fin antenna located on the measurement vehicle.
(c) Yagi antenna. (d) AP set on the wall 234

Figure 10.8   (a) Cost function J at each communication cycle
under the greedy policy and the proposed cognitive
control policy. (b) The cost function J at each
communication cycle under the SMDP policy, and the
proposed cognitive control policy 236

Figure 10.9   Train travel trajectory under the greedy policy
(the headway is 15 s) 237

Figure 10.10   Train travel trajectory under the SMDP policy
(the headway is 15 s) 237

Figure 10.11   Train travel trajectory under the proposed cognitive
control policy (the headway is 15 s) 238

Figure 10.12   Train travel trajectory under the proposed cognitive
control policy (the headway is 90 s) 238

Figure 10.13   Train travel trajectory under the SMDP policy
(the headway is 90 s) 239

Figure 10.14   Train travel trajectory under the greedy policy
(the headway is 90 s) 239

Figure 10.15   Handoff latency under different policies 240

Figure 10.16   Train–ground failure rate under different policies 241

Figure 10.17   Performance of optimization versus steps 243

List of Tables

Table 1.1 Radio-Based CBTC Projects around the World 9

Table 3.1 Advantages and Disadvantages of NDT Methods for Rail and Fastening Parts Inspection 58

Table 3.2 Comparison of NDT Methods for Rail and Fastening Parts 61

Table 4.1 Notions of Symbols 71

Table 4.2 Thresholds of SNR Levels (Four Levels) at the Location of 100 m for Different Intervals 76

Table 4.3 Thresholds of SNR Levels (Eight Levels) at the Location of 100 m for Different Intervals 77

Table 4.4 State Transition Probabilities of the FSMC Model and the Measurement Data with Four States and 5 m Interval at the Location (35–40 m) 77

Table 6.1 Parameters Used in Numerical Examples 110

Table 7.1 Main Notations 118

Table 7.2 Simulation Parameters 139

Table 8.1 Simulation Parameters 168

Table 9.1 Threshold of the Mean Envelope Power at Certain Possibilities ($\sigma_\Omega = 8$ dB) 189

Table 9.2 Threshold of the Envelope Power at Certain Probabilities 189

Table 9.3 Parameters for AP Coverage Plan 190

Table 9.4 Parameters to Analyze the Rate of Packet Drops Introduced by Handovers 193

Table 9.5 Field Test Results on the Rate of Packet Drops Introduced by Handovers 201

Table 9.6 Simulation Parameters 202

Table 9.7 LMI Feasibility of the System Using Current Control Scheme
($T = 0.3$ s) 204

Table 10.1 Availability under Different Policies...235

Table 10.2 Parameters Used in the Simulations...242

Preface

Introduction

With rapid population explosion, improving rail transit speed and capacity is strongly desirable around the world. Communications-based train control (CBTC) is an automated train control system using high capacity and bidirectional train–ground communications to ensure the safe operation of rail vehicles. As a modern successor to traditional railway signaling systems using track circuits, interlockings, and signals, CBTC can improve the capacity of railway network infrastructure and enhance the level of safety and service offered to customers.

CBTC systems have opened up several areas of research, which have been explored extensively and continue to attract research and development efforts. This book features some of the major advances in the research on CBTC systems. The contributed chapters in this book from leading experts in this field cover different aspects of modeling, analysis, design, testing, management, deployment, and optimization of algorithms, protocols, and architectures of CBTC systems. A summary of all of the chapters is provided in the following sections.

As the first chapter of this book, Chapter 1, authored by Li Zhu, F. Richard Yu, and Fei Wang, presents the background and evolution of train signaling/train control systems. Then it introduces the main features and architecture of CBTC systems. Some challenges of CBTC systems are presented. The chapter also describes the main CBTC projects around the world.

Chapter 2, authored by Kenneth Diemunsch, explains why transit agencies decide to use CBTC for new lines or for upgrading their signaling systems. Then, the author explains the reason for performing specific tests at the factory or in the field and provides insight based on experience with several CBTC projects in the last decade.

Chapter 3, authored by Vassilios Kappatos, Tat-Hean Gan, and Dimitris Stamatelos, discusses nondestructive testing techniques that can be employed to inspect rails and fastening parts as well as relevant research and development work in this field. As nondestructive testing techniques significantly depend on the nature of defects, a discussion about the defects that emerge on the rail infrastructure is included. Finally, an overview of the capacity of the recent train protection methods mainly based on the CBTC is carried out for a complete overview of all measures (nondestructive testing, train protection) that can be used to avoid any potential and serious rail accidents.

Channel Modeling

Chapter 4, authored by Hongwei Wang, F. Richard Yu, Li Zhu, and Tao Tang, develops a finite-state Markov channel (FSMC) model for tunnel channels in CBTC systems. The proposed FSMC model is based on real field CBTC channel measurements obtained from a business operating subway line. Unlike most existing channel models, which are not related to specific locations, the proposed FSMC channel model takes train locations into account to have a more accurate channel model. The distance between the transmitter and the receiver is divided into intervals, and an FSMC model is applied in each interval. The accuracy of the proposed FSMC model is illustrated by the simulation results generated from the model and the real field measurement results.

Chapter 5, authored by Hongwei Wang, F. Richard Yu, Li Zhu, and Tao Tang, discusses modeling the wireless channels with a leaky waveguide for CBTC systems. For viaduct scenarios, leaky rectangular waveguides are a popular approach, as they provide better performance and stronger anti-interference ability compared to free space. Based on the measurement results on the Beijing Subway Yizhuang Line, the authors use polynomial fitting and an equivalent magnetic dipole method to build the path loss model. In addition, the Akaike information criterion with a correction is applied to determine the distribution model of small-scale fading. The proposed path loss model of the channel with a leaky waveguide in CBTC systems is linear; the path loss exponent can be approximated by the transmission loss of the leaky waveguide. The authors show that small-scale fading follows a log-normal distribution, which is often referred to as the distribution model of small scale fading in body area communication propagation channels. In addition, the corresponding parameters of log-normal distribution are also determined from the measurement results.

Performance Analysis and Improvement with Advanced Communication Technologies

Chapter 6, authored by Li Zhu and F. Richard Yu, discusses the availability issue of WLAN-based data communication systems in CBTC. The authors propose two WLAN-based data communication systems with redundancy to improve availability in CBTC systems. The availability of WLAN-based data communication systems is analyzed using the continuous time Markov chain model. The transmission errors due to dynamic wireless channel fading and handoffs that take place when the train crosses the border of coverage areas of two successive access points are considered to be the main causes of system failures. The authors then model the WLAN-based data communication system behavior using the deterministic and stochastic petri net, which is a high-level description language for formally specifying complex systems. The deterministic and stochastic petri net solution is used to show the soundness of the proposed continuous time Markov chain model. The deterministic and stochastic petri net provides an intuitive and efficient way to describe complex system behavior and facilitates the modeling of system steady-state probability. Using numerical examples, the authors compare the availability of the two proposed WLAN-based data communication systems with that of an existing system that has no redundancy. The results show that the proposed data communication systems with redundancy have much higher availability compared to the existing system.

Chapter 7, authored by Li Zhu, F. Richard Yu, and Tao Tang, uses recent advances in coordinated multipoint transmission/reception (CoMP) to enable soft handoffs and consequently enhance the performance of CBTC systems. CoMP is a new method that helps with the implementation of dynamic base station coordination in practice. With CoMP, a train can communicate with a cluster of base stations simultaneously, a system that is different from the current CBTC systems, where a train can communicate with only a single base station at any given time. The authors jointly consider CoMP cluster selection and handoff decision issues in CBTC systems. In order to mitigate the impacts of communication latency on train control performance, they propose an optimal guidance trajectory calculation scheme in the train control procedure that takes full consideration of the tracking error caused by communication latency. Then, the system optimization of CBTC system with CoMP is formulated as a semi-Markov decision process. Extensive simulation results are presented to show that train control performance can be improved substantially in the proposed CBTC system with CoMP.

Chapter 8, authored by Hailin Jiang, Victor C. M. Leung, Chunhai Gao, and Tao Tang, proposes a multiple-input and multiple-output (MIMO)-assisted handoff scheme for CBTC WLAN systems with two or more antennas configured on the train and at each access point. A location-based handoff is proposed that takes advantage of the fact that the train in a CBTC system can acquire its locations accurately in real time. Information on train position from the mobile station is sent simultaneously with handoff signaling by means of MIMO multiplexing. The signaling and data packets are recovered at the access points by means of MIMO signal detection algorithms such as V-BLAST algorithms. The handoff performance, including the frame error rate of the handoff signaling, the handoff latency, the error-free period, and the impacts on the inter-site distance, is analyzed and compared with parameters associated with traditional handoff schemes.

Performance Analysis and Improvement with Advanced Control Technologies

Chapter 9, authored by Bing Bu, F. Richard Yu, and Tao Tang, integrates the design of trains' control and train–ground communication through modeling the control system of a group of trains in CBTC as a networked control system. The authors discuss packet drops in CBTC systems, introduce packet drops into the networked control system model, analyze their impact on the stability and performance of CBTC systems, and propose two novel control schemes to improve the performances of CBTC systems with random packet drops. Extensive field test and simulation results are presented to show that the proposed schemes can provide less energy consumption, better riding comfortability, and higher line capacity compared to existing scheme.

Chapter 10, authored by Hongwei Wang and F. Richard Yu, uses recent advances in cognitive dynamic systems in CBTC systems considering both train–ground communication and train control. In the cognitive control approach, the notion of information gap is adopted to quantitatively describe the effects of train–ground communication on train control performance. Specifically, as train–ground communication is used to exchange information between the train and control center, packet delay and drop lead to an information gap, which is the difference between the actual state and the observed state of the train. The wireless channel is modeled as finite-state Markov chains with multiple state transition probability matrices, which can demonstrate the characteristics of both large-scale fading and small-scale fading. The channel state transition probability matrices are derived from real field measurement results. Simulation results show that the proposed cognitive control approach can significantly improve train control performance in CBTC systems.

Conclusion

The chapters in this book essentially feature some of the major advances in the research on CBTC systems. Therefore, the book will be useful to both researchers and practitioners in this area. Readers will find the rich set of references in each chapter particularly valuable.

F. Richard Yu
Carleton University

Editor

F. **Richard Yu** earned his PhD degree in electrical engineering from the University of British Columbia in 2003. From 2002 to 2004, he was with Ericsson (in Lund, Sweden), where he worked on the research and development of wireless mobile systems. From 2005 to 2006, he was with a start-up in California, where he worked on research and development in the areas of advanced wireless communication technologies and new standards. He joined the Carleton School of Information Technology and the Department of Systems and Computer Engineering at Carleton University in 2007, where he is currently an associate professor. He received the IEEE Outstanding Leadership Award in 2013; Carleton Research Achievement Award in 2012; the Ontario Early Researcher Award (formerly Premier's Research Excellence Award) in 2011; the Excellent Contribution Award at IEEE/IFIP TrustCom in 2010; the Leadership Opportunity Fund Award from the Canada Foundation of Innovation in 2009; and best-paper awards at IEEE ICC 2014, Globecom 2012, IEEE/IFIP TrustCom 2009, and the International Conference on Networking 2005. His research interests include cross-layer/cross-system design, security, green IT, and QoS provisioning in wireless-based systems.

Dr. Yu serves on the editorial board of several journals; he is co-editor-in-chief for *Ad Hoc & Sensor Wireless Networks*; lead series editor for *IEEE Transactions on Vehicular Technology, IEEE Communications Surveys & Tutorials, EURASIP Journal on Wireless Communications Networking, Wiley Journal on Security and Communication Networks*, and *International Journal of Wireless Communications and Networking*; a guest editor for *IEEE Transactions on Emerging Topics in Computing* special issue *Advances in Mobile Cloud Computing*; and a guest editor for *IEEE Systems Journal* for the special issue *Smart Grid Communications Systems*. He has served on the Technical Program Committee of numerous conferences, as the TPC co-chair of IEEE GreenCom'14, INFOCOM-MCV'15, Globecom'14, WiVEC'14, INFOCOM-MCC'14, Globecom'13, GreenCom'13, CCNC'13, INFOCOM-CCSES'12, ICC-GCN'12, VTC'12S, Globecom'11, INFOCOM-GCN'11, INFOCOM-CWCN'10, IEEE IWCMC'09, VTC'08F, and WiN-ITS'07; as the publication chair of ICST QShine'10; and the co-chair of ICUMT-CWCN'09. Dr. Yu is a registered professional engineer in the province of Ontario, Canada.

Contributors

Bing Bu
State Key Laboratory of Rail Traffic
    Control and Safety
Beijing Jiaotong University
Beijing, China

Kenneth Diemunsch
CH2M HILL—Transit & Rail
New York, New York

Tat-Hean Gan
Brunel Innovation Centre
Brunel University
Uxbridge, United Kingdom

Chunhai Gao
State Key Laboratory of Rail Traffic
    Control and Safety
Beijing Jiaotong University
Beijing, China

Hailin Jiang
State Key Laboratory of Rail Traffic
    Control and Safety
Beijing Jiaotong University
Beijing, China

Vassilios Kappatos
Brunel Innovation Centre
Brunel University
Uxbridge, United Kingdom

Victor C. M. Leung
Department of Electrical and
    Computer Engineering
The University of British Columbia
Vancouver, British Columbia, Canada

Dimitris Stamatelos
Department of Aeronautical Sciences,
    Aeronautical Engineering Section
Hellenic Air Force Academy
Dekelia Air Force Base
Attica, Greece

Tao Tang
State Key Laboratory of Rail Traffic
    Control and Safety
Beijing Jiaotong University
Beijing, China

Fei Wang
State Key Laboratory of Rail Traffic
 Control and Safety
Beijing Jiaotong University
Beijing, China

Hongwei Wang
National Railway Safety Assessment
 Research Center
Beijing Jiaotong University
Beijing, China

Li Zhu
State Key Laboratory of Rail Traffic
 Control and Safety
Beijing Jiaotong University
Beijing, China

Chapter 1

Introduction to Communications-Based Train Control

Li Zhu, F. Richard Yu, and Fei Wang

Contents

1.1 Introduction...1
1.2 Evolution of Train Signaling/Train Control Systems....................................2
1.3 Main Features and Architecture of CBTC Systems......................................5
1.4 Challenges of CBTC Systems...7
1.5 Projects of CBTC Systems...8
1.6 Conclusion...13
References...13

1.1 Introduction

Rapid population explosion has resulted in a series of problems, such as traffic jam, environment pollution, and energy crisis. Recently, there has been a strong desire around the world to improve the rail transit speed and capacity in order to relieve the pressures from already-busy roads to address the need for fast, punctual, and environmentally friendly mass transit systems.

The train signaling systems need to evolve and adapt to safely meet this increase in demand and traffic capacity [1]. The main objective of communications-based train control (CBTC) signaling system is to increase the capacity by safely reducing the time interval (headway) between trains traveling along the line. Specifically, CBTC makes use of the communications between the railway track equipment and the train for train control and traffic management. Because the exact position of a train is known more accurately than with the traditional signaling system, the railway traffic can be managed more efficiently and safely.

As defined in the IEEE 1474 standard [2], a CBTC system is a "continuous, automatic train control system utilizing high-resolution train location determination, independent of track circuits; continuous, high-capacity, bidirectional train-to-wayside data communications; and trainborne and wayside processors capable of implementing automatic train protection (ATP) functions, as well as optional automatic train operation (ATO) and automatic train supervision (ATS) functions."

In this chapter, we first present the background and evolution of train signaling/train control systems. Then, we introduce CBTC systems, followed by the main CBTC projects around the world.

1.2 Evolution of Train Signaling/Train Control Systems

The main objective of a train signaling/train control system is to prevent collisions when trains travel on the railway track. Therefore, a common ingredient of various types of train signaling systems is as follows: the locations of the trains must be known by the system at some level of granularity.

The first generation of train control architecture includes track circuits for train detection, wayside signals to provide movement authority indications to train operators, and trip stops to enforce a train stop [1]. Figure 1.1 illustrates this architecture. In Figure 1.1, if track circuit TC5 is occupied (shunted by a train), the signal at the entrance to TC5 displays a red aspect. If block TC3 is unoccupied and TC5 is occupied, the entrance signal to TC3 displays a yellow aspect. If both TC1 and TC3 are unoccupied, the entrance signal to TC1 displays a green aspect. These signals are separated by the train's safe braking distance (SBD), which is calculated and set at a sufficient length for a train to stop safely from the maximum operating speed specified for the track section. We can see that, in this system, a green aspect means that two blocks (or at least twice SBD) are clear ahead of the signal; a yellow aspect means that one block (at least SBD) is clear ahead of the signal; and a red aspect means that the block ahead has a train occupying the track circuit.

Figure 1.1 Train signaling system using wayside signals.

This simple train signaling system is similar to road traffic light signaling systems. Due to its simplicity, this train control philosophy has served the industry well and continues in service operation at many major train transit systems around the world. Nevertheless, with this train signaling architecture, the wayside logic does not know the exact location of the train in the track circuit. Instead, it only knows that the train is located somewhere in the block. Because the blocks cannot move, this kind of systems is known as fixed block systems. The main drawback of this system is that the achievable train throughput and operational flexibility are limited by the fixed-block, track circuit configuration and associated wayside signal aspects.

The next evolution in train signaling was also track circuit-based with the wayside signals replaced by in-cab signals, providing continuous ATP through the use of speed codes transmitted from the wayside through the running rails to the train. Such coded track circuits were developed by signaling suppliers in the United States around the middle of the last century. Although they were not immediately applied to transit railways, they ultimately made a significant contribution to the next-generation train control systems. With this train control architecture, a portion of the train control logic and equipment is transferred to the train, with equipment capable of detecting and reacting to speed codes, and displaying movement authority information (permitted speed and signal aspects) to the train operator. This generation of train control technology permits automatic driving modes, but train throughput and operational flexibility are still limited by the track circuit layout and the number of available speed codes. This generation of signaling technology entered service in the latter half of the twentieth century, including the Washington (WMATA), Atlanta (MARTA), and San Francisco (BART) systems in the United States, the London Underground's Victoria Line, and the initial rail lines in Hong Kong and Singapore. Many rail transit agencies also adopted this technology in order to transit ATO with continuous ATP, such as London Underground's Central Line resignaling.

The next significant evolution in the train signaling architecture continued the trend to provide more precise control of train movements by increasing the amount of data transmitted to the train such that the train could now be controlled and supervised to follow a specific speed–distance profile, rather than simply responding to a limited number of individual speed codes. This generation of train control technology, also referred to as "distance-to-go" technology, can support automatic driving modes and improve train throughput. Under this train control architecture, the limits of a train's movement authority are still determined by track circuit occupancies, as illustrated in Figure 1.2.

The wayside processor in Figure 1.2, knowing the location of all trains via track circuits, can generate coded messages to each track circuit. This information contains the permitted line speed, the target speed for the train, and the distance-to-go to the target speed. Using this information, the train's onboard equipment calculates the speed–distance profile for the train to follow. In addition, a track map database with grade, curvature, station location, and civil speed limit information is stored within each train's cab signaling equipment. The train knows which track circuit it

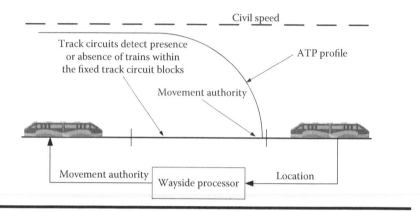

Figure 1.2 Profile-based train control system.

is in via a unique ID of the cab signaling information. The cab signaling equipment then uses the track map database to calculate the accurate speed–distance profile.

The next generation of the train control architecture is generally referred to as CBTC. The goal of a CBTC system is the same as the traditional systems, for example, safe train separation; however, it also has the challenge of minimizing the amount of wayside and trackside equipment. This means the elimination of traditional train detection devices, that is, track circuits. Similar to the previous generations of train control technologies, CBTC supports automatic driving modes and controls/supervises train movements in accordance with a defined speed–distance profile. For CBTC systems, however, movement authority limits are no longer constrained by physical track circuit boundaries but are established through train position reports that can provide for "virtual block" or "moving block" control philosophies, as illustrated in Figure 1.3.

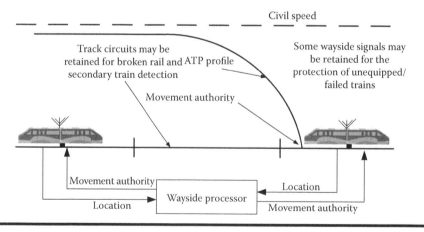

Figure 1.3 CBTC system.

In CBTC systems, a major portion of the train control logic is now located within the train-borne CBTC equipment, and a geographically continuous train-to-wayside and wayside-to-train data communications network permits the transfer of significantly more control and status information than is possible with the earlier generation train control systems. As such, CBTC systems offer the greatest operational flexibility and can support the maximum train throughput, constrained only by the performance of the rolling stock and the limitations of the physical track alignment. In particular, the high level of control provided by CBTC systems makes this the technology of choice for driverless/unattended train operations (UTOs).

1.3 Main Features and Architecture of CBTC Systems

With the exact location information of a train in CBTC systems, the following train can follow up the rear of the train with a moving block system. Specifically, the train location and its braking curve are continuously calculated by the trains, and then communicated to the wayside equipment. Then, the wayside equipment is able to establish protected areas, each one called limit of movement authority (LMA). In addition, CBTC systems use closed-loop control between the train and the ground control center to improve the reliability of train control. Consequently, this results in a reduced headway between consecutive trains and an increased transport capacity. Moreover, using digital radio transmissions, CBTC can achieve two-way large-capacity communications between the train and the wayside, which can reduce unnecessary train acceleration and deceleration braking, improve passenger comfort, and enable significant energy savings.

CBTC systems allow different levels of automation or Grades of Automation (GoA), as defined and classified in IEC 62290-1 [3]. The grades of automation available range from a manual protected operation, GoA 1 (usually applied as a fallback operation mode), to the fully automated operation, GoA 4 (UTO). Intermediate operation modes comprise semiautomated GoA 2 (semiautomated operation [STO] mode) or driverless GoA 3 (driverless train operation [DTO]). The latter operates without a driver in the cabin but requires an attendant to face degraded modes of operation as well as guides the passengers in the case of emergencies. Please note that, although CBTC systems are considered as a basic technology for "driverless" or "automated trains," it is not a synonym for them.

The typical architecture of a modern CBTC system consists of the following main components, as shown in Figure 1.4.

1. *Wayside equipment.* It includes the interlocking and the subsystems controlling every zone in the line or network (typically containing the wayside ATP and ATO functionalities). The control of the system is performed from a command ATS, though local control subsystems may also be included. Depending on the suppliers, the architecture may be centralized or distributed.

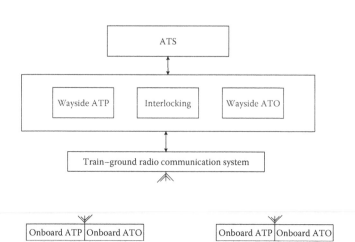

Figure 1.4   Typical architecture of a modern CBTC system.

   − *ATS.* This system is commonly integrated within most of the CBTC solutions. Its main task is to act as the interface between the operator and the system, managing the traffic according to the specific regulation criteria. Other tasks may include the event and alarm management as well as act as the interface with external systems.
   − *Interlocking.* When needed as an independent subsystem (for instance, as a fallback system), it will be in charge of the vital control of the trackside objects such as switches or signals, as well as other related functionality. In the case of simpler networks or lines, the functionality of the interlocking may be integrated into the wayside ATP system.
   − *Wayside ATP.* This subsystem undertakes the management of all the communications with the trains in its area. Additionally, it calculates the limits of movement authority (LMAs) that every train must respect while operating in the mentioned area. This task is therefore critical for the operation safety.
   − *Wayside ATO.* It is in charge of controlling the destination and regulation targets of every train. Its functionality provides all the trains in the system with their destination as well as with other data such as the dwell time in the stations. Additionally, it may also perform auxiliary and non-safety-related tasks including, for instance, alarm/event communication and management, or handling skip/hold station commands.
 2. *CBTC onboard equipment.* It includes ATP and ATO subsystems in the vehicles.
   − *Onboard ATP.* This subsystem is in charge of the continuous control of the train speed according to the safety profile and applying the brake if it is necessary. It is also in charge of the communication with the wayside ATP subsystem in order to exchange the information needed for a safe

operation (sending speed and braking distance and receiving the LMA for a safe operation).

–   *Onboard ATO*. It is responsible for the automatic control of the traction and braking effort in order to keep the train under the threshold established by the ATP subsystem. Its main task is either to facilitate the driver or attendant functions or even to operate the train in a fully automatic mode while maintaining the traffic regulation targets and passenger comfort. It also allows the selection of different automatic driving strategies to adapt the run time or even reduce the power consumption.

3. *Train–ground radio communication subsystem*. It is one of the key technologies in CBTC systems. Wireless networks, such as Global System for Mobile Communications—Railway and wireless local area networks (WLANs), are commonly used to provide bidirectional train–ground communications. For urban mass transit systems, IEEE 802.11a/b/g-based WLANs are a better choice due to the available commercial off-the-shelf equipment and the philosophy of open standards and interoperability.

1.4 Challenges of CBTC Systems

Although there are many advantages of CBTC systems, several significant research challenges remain to be addressed to make CBTC systems safer, more reliable, and efficient.

The primary challenge of a CBTC system is that if the radio communications link between any of the trains is disrupted, then all or part of the system might have to enter a fail-safe state until the problem is remedied. Communications failures can result from equipment malfunction, electromagnetic interference, weak signal strength, frequent handoff, or saturation of the communications medium. Building a train control system over wireless networks is a challenging task. Due to unreliable wireless communications and train mobility, the train control performance can be significantly affected by wireless networks. This is the reason why, historically, CBTC systems first implemented radio communication systems in 2003, when the required technology was mature enough for critical applications.

Depending on the severity of the communication loss, this state can range from vehicles temporarily reducing speed, coming to a halt, or operating in a degraded mode until communications are reestablished. If communication outage is permanent, some sort of contingency operation must be implemented, which may consist of manual operation using absolute block or, in the worst case, the substitution of an alternative form of transportation. As a result, high availability of CBTC systems is crucial for proper operation, especially if we consider that such systems are used to increase transport capacity and reduce headway. System redundancy and recovery mechanisms must then be thoroughly checked to achieve a high robustness in operation. With the increased availability of the CBTC system, the need for an extensive training and periodical refresh of system operators on the recovery procedures must also be considered.

With the emerging services over open industrial, scientific, and medical radio bands (i.e., 2.4 and 5.8 GHz) and the potential disruption over critical CBTC services, there is an increasing pressure in the international community to reserve a frequency band specifically for radio-based urban rail systems. Such decision would help standardize CBTC systems across the market and ensure availability for those critical systems.

Another challenge lies in systems with poor line-of-sight or spectrum/bandwidth limitations. A larger than anticipated number of transponders may be required to enhance the service. This is usually more of an issue with applying CBTC to existing transit systems in tunnels that were not designed from the outset to support it. An alternate method to improve system availability in tunnels is the use of leaky feeder cable that, while having higher initial costs, achieves a more reliable radio link.

As a CBTC system is required to have high availability and, particularly, allows for a graceful degradation, a secondary method of signaling might be provided to ensure some level of nondegraded service upon partial or complete CBTC unavailability. This is particularly relevant for brownfield implementations (lines with an already existing signaling system) where the infrastructure design cannot be controlled and the coexistence with legacy systems is required, at least, temporarily.

Security is another big concern in CBTC systems. There are many risks in CBTC systems needed to be considered seriously due to the distinctive features of CBTC, including open wireless transmission medium, nomadic trains, and lack of dedicated infrastructure of security protection. Therefore, in addition to the vulnerabilities and threats of traditional wireless-based systems, the involvement of intelligence in CBTC presents new security challenges. For many security issues, authentication is an important requirement, which is crucial for integrity, confidentiality, and nonrepudiation. In addition, the experience in security of traditional wired and wireless networks indicates the importance of multilevel protections because there are always some weak points in the system, no matter what is used for prevention-based approaches (e.g., authentication). This is especially true for CBTC systems, given the low physical security autonomous functions of trains. To solve this problem, detection-based approaches [e.g., intrusion detection systems (IDSs)], serving as the second wall of protection, can effectively help identify malicious activities. Both prevention-based and detection-based approaches need to be carefully studied for CBTC systems.

In addition, there is the probability of human error and improper application of recovery procedures if the system becomes unavailable. Therefore, it is important to enhance the operator's safety education and training, ensuring safe operation of trains.

1.5 Projects of CBTC Systems

CBTC technology has been (and is being) successfully implemented for a variety of applications. Table 1.1 summarizes the main radio-based CBTC systems deployed around the world as well as those ongoing projects being developed [4]. They range

Table 1.1 Radio-Based CBTC Projects around the World

Country	Location	Line/System	Supplier	Solution	Commissioning	LoA
USA	Washington Dulles Airport	Dulles APM	Thales	SelTrac	2009	UTO
	Seattle–Tacoma Airport	Satellite Transit System APM	Bombardier	CITYFLO 650	2003	UTO
	Sacramento International Airport	Sacramento APM	Bombardier	CITYFLO 650	2011	UTO
	Massachusetts Bay Transportation Authority	Ashmont–Mattapan High Speed Line	Argenia	SafeNet CBTC	2014	STO
Canada	Toronto Metro	YUS line	Alstom	Urbalis	2013	STO
	Ottawa Light Rail	Confederation Line	Thales	SelTrac	2018	STO
	Edmonton Light Rail Transit	Capital Line Metro Line	Thales	SelTrac	December 2014	DTO
China	Beijing Metro	1, 2, 6, 9	Alstom	Urbalis	From 2008 to 2015	STO
	Beijing Metro	4	Thales	SelTrac	2009	STO

(Continued)

Table 1.1 (*Continued*)   Radio-Based CBTC Projects around the World

Country	Location	Line/System	Supplier	Solution	Commissioning	LoA
	Beijing Metro	8, 10	Siemens	Trainguard MT CBTC	2013	STO
	Wuhan Metro	2, 4	Alstom	Urbalis	2013	STO
	Wuxi Metro	1, 2	Alstom	Urbalis	2015	STO
	Tianjin Metro	2, 3	Bombardier	CITYFLO 650	2012	STO
	Shanghai Metro	10, 12, 13, 16	Alstom	Urbalis	From 2010 to 2013	UTO and STO
	Nanjing Metro	Nanjing Airport Rail Link	Thales	SelTrac	2014	STO
	Kunming Metro	1, 2	Alstom	Urbalis	2013	
	Hong Kong MTRC	Hong Kong APM	Thales	SelTrac	2014	UTO
	Guangzhou Metro	6	Alstom	Urbalis	2012	ATO
	Taipei Metro	Circular	Ansaldo STS	CBTC	2015	UTO
Singapore	Singapore Metro	North South Line	Thales	SelTrac	2015	UTO
	Singapore Metro	Downtown Line	Invensys	Sirius	2016	UTO

(Continued)

Table 1.1 (Continued)   Radio-Based CBTC Projects around the World

Country	Location	Line/System	Supplier	Solution	Commissioning	LoA
Malaysia	Kuala Lumpur Rail Transit	Ampang Line	Thales	SelTrac	2016	UTO
	Kuala Lumpur MRT	Klang Valley MRT	Bombardier	CITYFLO 650	2017	UTO
Korea	Seoul Metro	Bundang Line	Thales	SelTrac	2014	UTO
	Incheon Metro	2	Thales	SelTrac	2014	UTO
Saudi Arabia	Riyadh	KAFD Monorail	Bombardier	CITYFLO 650	2012	UTO
	Dubai Metro	Red, Green	Thales	SelTrac	2011	UTO
	Dubai Metro	Al Sufouh LRT	Alstom	Urbalis	2014	STO
India	Hyderabad Metro Rail	L1, L2, L3	Thales	SelTrac	2016	UTO
	Delhi Metro	Line 7	Bombardier	CITYFLO 650	2015	
France	Paris Metro	1	Siemens	Trainguard MT CBTC	2011	DTO
	Paris Metro	13	Thales	SelTrac	2013	STO

(Continued)

Table 1.1 (*Continued*) Radio-Based CBTC Projects around the World

Country	Location	Line/System	Supplier	Solution	Commissioning	LoA
	London Gatwick Airport	Terminal Transfer APM	Bombardier	CITYFLO 650	2010	UTO
	Lille Metro	1	Alstom	Urbalis	2017	UTO
Spain	Málaga Metro	1, 2	Alstom	Urbalis	2013	
	Metro de Madrid	7 Extension MetroEste	Invensys	Sirius	2011	STO
Brazil	Sao Paulo Metro	5	Bombardier	CITYFLO 650	2015	UTO
	Sao Paulo Metro	17	Thales	SelTrac	2015	UTO
Mexico	Mexico City Metro	12	Alstom	Urbalis	2012	STO
Switzerland	Lausanne Metro	M2	Alstom	Urbalis	2008	UTO
Finland	Helsinki Metro	1	Siemens	Trainguard MT CBTC	2014	STO
Denmark	Copenhagen S-Train	All lines	Siemens	Trainguard MT CBTC	2018	STO
Venezuela	Caracas Metro	1	Invensys	Sirius	2013	
Hungary	Budapest Metro 2014 (M4)	M2, M4	Siemens	Trainguard MT CBTC	2013 (M2)	

APM, automated people mover; LoA, level of automation; STO, semiautomated operation mode; UTO, unattended train operation.

from some implementations with short track, limited numbers of vehicles, and few operating modes (such as the airport automated people movers in San Francisco or Washington), to complex overlays on existing railway networks carrying more than a million passengers each day and with more than 100 trains (such as lines 1 and 6 in Metro de Madrid, line 3 in Shenzhen Metro, some lines in Paris Metro and Beijing Metro, or the Sub-Surface Railway (SSR) in London Underground).

1.6 Conclusion

As a modern successor of traditional railway signaling systems using track circuits, interlockings, and signals, CBTC is an automated train control system using high-capacity and bidirectional train-ground communications to ensure the safe operation of rail vehicles, improve the utilization of railway network infrastructure, and enhance the level of service offered to customers. In this chapter, we introduced several traditional train control systems. Then, we presented the main features and architecture of CBTC systems. Several significant research challenges of CBTC systems were discussed. Finally, we summarized the main CBTC projects around the world.

References

1. R. D. Pascoe and T. N. Eichorn. What is communication-based train control? *IEEE Veh. Tech. Magazine*, 4(4):16–21, 2009.
2. *IEEE. IEEE Std 1474.1-2004:* IEEE standard for communications-based train control (CBTC) performance and functional requirements *(Revision of IEEE Std 1474.1-1999). IEEE,* pp. 1–45, 2004.
3. Railway applications—Urban guided transport management and command/control systems—Part 1: System principles and fundamental concepts. *IEC 62290-1,* 2006.
4. CBTC. CBTC projects. *www.tsd.org/cbtc/projects.* Accessed September 2014.

Chapter 2

Testing Communications-Based Train Control

Kenneth Diemunsch

Contents

2.1 Introduction 16
2.2 Pros and Cons of Using CBTC System 18
   2.2.1 Pros of CBTC 18
   2.2.2 Cons of CBTC 19
2.3 Different Types of CBTC Projects..19
   2.3.1 Installation on a New Line: Greenfield Project 19
   2.3.2 Migrating an Existing Line: Brownfield Project................20
2.4 CBTC Architecture...21
2.5 Principles of CBTC Testing 21
   2.5.1 Reuse as Much as Possible from Previous Projects............21
   2.5.2 Test in Factory as Much as Possible 22
   2.5.3 Test All Safety-Related Items 22
2.6 Environmental Tests..23
   2.6.1 EMC Tests..23
   2.6.2 Climatic Conditions 23
   2.6.3 Mechanical Conditions...24
   2.6.4 Abrasive Conditions..24
2.7 First Article Configuration Inspection 24
2.8 Factory Tests 24
   2.8.1 Factory Test Goals..25
   2.8.2 Factory Setup...25
   2.8.3 Different Types of Factory Tests 26

      2.8.3.1   Product Factory Tests 26

      2.8.3.2   CBTC Supplier Internal Factory Testing...............26

      2.8.3.3   Factory Acceptance Test 26

      2.8.3.4   Description of the Tests to Be Performed 26

      2.8.3.5   Major Challenges of Factory Tests........................27

2.9   On-Board Integration Tests..27

    2.9.1   Rolling Stock Characterization Tests.....................................27

    2.9.2   Mechanical and Electrical Tests 28

    2.9.3   Static and Dynamic Post Installation Check Out Tests 28

2.10   Test Track 28

    2.10.1 Use of the Test Track..28

    2.10.2 Test Track Equipment 29

    2.10.3 Location of the Test Track..29

2.11   On-Site Tests...30

    2.11.1 Post Installation Check Out 30

    2.11.2 DCS Tests 30

      2.11.2.1 Wayside Network Tests 30

      2.11.2.2 Radio Tests..31

    2.11.3 Localization Tests..32

    2.11.4 Integration Tests...32

      2.11.4.1 Integration Tests: Internal to CBTC.....................32

      2.11.4.2 Integration Tests: External to CBTC.....................33

    2.11.5 Functional Tests..33

    2.11.6 ATO Tests...35

    2.11.7 ATS Tests..35

    2.11.8 Site Acceptance Tests...35

    2.11.9 Shadow Mode Tests 36

    2.11.10 Reliability, Availability, and Maintenance Tests......................36

      2.11.10.1 Reliability and Availability Demonstration............36

      2.11.10.2 Maintainability Demonstration............................37

2.12   CBTC Test Duration 38

2.13   Constraints on Field Tests 38

2.14   Conclusion 39

References 40

2.1 Introduction

Communications-based train control (CBTC) technology is the most advanced train control system for urban railway infrastructures. It is very different from conventional relay-based signaling systems and more complex than most cab signaling systems. CBTC functions are numerous and highly complex with customized details for each project. They cannot be tested for all possible conditions at all locations. Knowledge of the CBTC system and experience with train control commissioning are key to performing enough tests to detect most issues but permit the start of revenue service as early as possible. The testing strategy proposed by the CBTC supplier is the result of years of experience with the goal of minimizing expensive field tests while demonstrating that the system will work properly in revenue service. Despite the numerous tests performed before revenue service, it is inevitable that operating challenges will be faced during the first months of CBTC system operation.

CBTC suppliers have CBTC products corresponding to a specific system architecture with core functions that have been tested and operating in revenue service on many transit properties. Most transit agencies require a service-proven technology that they want to customize. The level of customization of the CBTC system for the project is one of the main factors in the number of issues in the system. Inevitably, on every CBTC project, customized functions are where most of the errors are found. They may be related to design or implementation. Some customizations are inevitable, for instance, fitment of the carborne equipment onto the train, or for taking advantage of a new improved functionality of the product. CBTC suppliers prefer to deploy a system as close as possible to their product which has been intensively tested in previous projects. It is common that a transit agency insists on buying an off-the-shelf CBTC system but ends up requiring many customized functions. Understanding that the system is a proven technology customized for the project is key to optimize the tests.

Testing of CBTC is done intensively in the factory. The factory setup allows for testing of almost all functions and all situations. The functions that may not be completely testable in the factory are related to the carborne controller (CC) interfaces and to the field characteristics such as train localization and radio coverage. Apart from those items, in an ideal world, the tests in the field would only be for demonstration to the transit agency that the system meets the contract requirements.

Today, most of the current CBTC projects are migration projects with the goal to increase revenue service performance by replacing an existing signaling system that has reached the end of life with the state-of-the-art CBTC technology. In almost all upgrade projects, the transit agency requires that the transition to CBTC system be performed with the least amount of impact on train service. This constraint increases the complexity of the project due to limited track access. In those cases, CBTC testing must be optimized to the extreme in order to be able to deploy the system while maintaining service.

The recent IEEE Std 1474.4-2011 Recommended Practice for Functional Testing of a Communications-Based Train Control (CBTC) System [1] provides a good reference and describes how and where CBTC functions should be tested. However, it does not describe the sequence of tests in the context of a project where CBTC is deployed on a transit property. This chapter intends to explain the reason for performing specific tests at the factory or in the field and to provide insight based on experience with several CBTC projects in the last decade.

2.2 Pros and Cons of Using CBTC System

Before presenting the testing strategy, let us consider the reason why transit agencies decide to use CBTC for new lines or for upgrading their signaling system.

2.2.1 Pros of CBTC

The main reasons for a transit agency to select CBTC technology are:

- *Safety.* CBTC includes continuous automatic train protection (ATP). Many conventional signaling systems enforce the speed of the train only at certain locations using grade time signals. After the train engineer accepts such signal, the train may then be operated to its maximum speed until the next red signal. This results in intermittent speed control that relies on human intervention not supervised by any system.
- *Throughput.* CBTC uses a moving or fixed virtual block system that allows trains to follow each other very closely resulting in improved headway performance. For projects with track circuits, more than one CBTC-equipped train can occupy an individual track circuit at any time, whereas conventional signaling system requires one or several track circuits between trains. In busy transit systems, a goal of CBTC migration is to improve minimum headway between trains.
- *Automatic train operation (ATO).* The CBTC system includes an ATO mode that enables the CC to control the movement of the train without the train engineer controlling the rolling stock master controller. The ATO mode provides a smoother ride for passengers, results in more predictable operation, and enables energy saving. ATO operation can be associated with driverless operation for additional benefits. ATO is also possible with other types of train control implementations.
- *Positive train control (PTC) compliant.* CBTC is a type of PTC, including work zone and slow speed order enforcement that comply with the Railway Safety Improvement Act of 2008 mandated by the U.S. Congress. Therefore, CBTC is a potential PTC solution for U.S. railroads under the jurisdiction of the Federal Railroad Administration.
- *High system availability.* CBTC includes redundancy and built-in diagnostic systems which report the status of most equipment to the automatic train supervision (ATS). These functions are also possible with other types of train control implementations.

- *Reduced maintenance cost.* CBTC has less equipment than conventional signaling systems, in particular on the trackside.
- CBTC has been deployed for more 30 years all over the world and is now a proven technology.

2.2.2 Cons of CBTC

CBTC technology also comes with several challenges for transit agencies:

- Initial investment cost of deploying the technology may be higher than other types of train control. Design, hardware, installation, and testing of CBTC require years of effort and usually take longer than expected.
- The system is not modifiable by the transit agency for different reasons. The first reason is technical: the computer skills required may not be available within the agency. The second reason is the responsibility: the original equipment manufacturer (OEM) is liable to provide a safe system to the transit agency. Transit agencies do not want to accept the responsibility for changing such complex system. Any change has to be carried out by the OEM. Note that there are efforts by large transit agencies such as New York City Transit and Régie Autonome des Transports Parisiens to specify interoperability and compatibility of products across multiple CBTC vendors.
- CBTC technology is very different from traditional relay-based signaling system, and, therefore, the transit agency must adapt to it. It requires new skills for engineering and maintenance personnel as well as a new organization in the transit agency.

2.3 Different Types of CBTC Projects

2.3.1 Installation on a New Line: Greenfield Project

The term "greenfield project" refers to the case where a transit agency is building a new line. The railway project involves all aspects of railway engineering such as civil engineering, track installation, traction power system, Rolling Stock and the signaling system. In the early 1980s, CBTC started to be implemented on such projects. Only after the technology was considered matured and experience was sufficient, CBTC was applied to lines already in service as described in Section 2.3.2.

Though new line schedules are more and more challenging, testing the signaling system on a greenfield project is facilitated with easier track access than on a project where the line is used for passenger service. Flexibility of track access is not the only advantage of testing on greenfield projects, but it is by far the most important. Access can be more frequent and reorganized at the last minute depending on previous tests and software development. There is no transportation department to interface with during testing, which is usually a very time-consuming interface. Another advantage is that the interfaced system may also be under development so interface design issues can be resolved by changing the CBTC or the other systems. New lines are often for new or recent transit systems which are more likely to embrace new technology and accept the method of testing recommended by the CBTC suppliers.

The disadvantage of greenfield projects is that the signaling system is the last part of a complete transportation system dependent on other systems. The result is that the signaling project is planned with a very compressed schedule without any slack in order to absorb previous system delays. Another difficulty for projects on new lines is that the other systems, which also require track access, may not be working properly during the CBTC tests. In particular, the rolling stock may be ready for CBTC tests but not for revenue operations, and the remaining rolling stock issues may affect CBTC tests if a train becomes stranded during testing.

2.3.2 Migrating an Existing Line: Brownfield Project

In large cities where the railway infrastructure was created decades ago, the cost of building a new line is very high and the time to create a new line is very long. Transit agencies prefer improving the capacity of existing lines in only a few years. Buying new trains with better performance and more passenger space is a possibility to improve the line capacity. In addition to or instead of buying new trains, the transit agency might decide to upgrade to CBTC technology to increase capacity by minimizing the headway between trains. Upgrading the signaling system of an existing line is referred as a brownfield project.

On a migration project, the biggest challenge is to get sufficient track access to install and test the new system while maintaining revenue service operation [2,3]. During revenue service hours, tracks are used to transport passengers, and during off hours or during nonpeak hours, maintenance actions to support the revenue service operation are performed. Installation and testing must be integrated with the maintenance schedule of the existing transportation system for the duration of the field activities of the CBTC project.

Transit agencies may plan to convert only one line or their entire system to CBTC. Several migration approaches, which affect testing strategy, have been used. One approach is to deploy CBTC on a line that is not very busy in order to learn about the system and to minimize the risk of deploying CBTC on a more busy line, usually under strong political scrutiny. Another approach is to directly use CBTC on the busiest line because the capacity needs to be improved as quickly as possible. Those brownfield projects on busy lines have the highest planning risk and need both an experienced CBTC supplier and a transit agency familiar with CBTC. Some of the very ambitious projects have failed or were considerably delayed, lasting more than 10 years. Finally, another approach is to have mega projects where all lines are upgraded at the same time using one or several suppliers, where test results on the first line may be used for the other lines to cut the total test time.

2.4 CBTC Architecture

The CBTC system considered in the this chapter is described in Refs. [4–6]. It is composed of four subsystems:

- *Carborne Controller (CC)*. The CC, also called OnBoard Controller Unit (OBCU), is located on board the train. It is responsible for determining the train speed, the train location, and the enforcement of the speed limit based on the movement authority limit (MAL).
- *Zone controller (ZC)*. The ZCs are located in the technical rooms. They are responsible for calculating and providing the MAL to the CC based on the information received from the trains and from other subsystems such as the interlocking. There are several ZCs per project to provide coverage for the full line. They exchange information with the onboard controllers in their territory and controll, directly or through an external signal system, field equipment such as switches.
- *ATS system*. The ATS system regulates train operations. It includes a human–machine interface with the operators at the Operations Control Center (OCC).
- *Data communication system (DCS)*. The DCS includes both the wayside communication network and the train to wayside communication system.

2.5 Principles of CBTC Testing

2.5.1 Reuse as Much as Possible from Previous Projects

There are only a few CBTC suppliers around the world, and most of them have been developing CBTC technology over the past decades. CBTC suppliers have different CBTC products depending on the features such as driverless operation. Their systems are the result of improvements over the years using feedback from previous projects regarding train operation, lessons learned on the deployment of the system, and advances in technology such as the IEEE 802.11 standard WiFi [7]. When a new product is developed, CBTC suppliers reuse the current version of their product, both the hardware and the software, and include improvements. The new CBTC product is then deployed on several transit properties in parallel until the next generation. It is common that the new product is developed for a large project identified as a development project. Transit agencies want a service-proven technology, which takes into account the latest technology with plans to customize it.

This strategy of using the same product for several projects has two direct impacts on the tests of CBTC:

- The first project of a new CBTC product is much more difficult than the previous projects done by the supplier. After the first project, other projects benefit from the work already done, and only specific features of the following projects are challenges.
- Previous development and test results can be reused. Because CBTC is a very complex and large system, reusing as much as possible the same hardware and software while adding improvement is a must. Core functions of CBTC have been the same for decades. The product team of the CBTC suppliers is in charge of performing low-level tests transparent to the transit agency. Items that can be reused are not only the software of core functions but also the factory setup, including simulators, and documentation, such as test procedure or safety analysis. Transit agencies expect that the CBTC supplier reuses previous projects to develop the system on their property, but test results from another projects cannot be used. The use of other project is valid for CBTC supplier internal purposes only.

2.5.2 Test in Factory as Much as Possible

In order to minimize field tests that require track access with far more resources than factory tests, a complete set of factory tests must be performed before using the software on-site. CBTC systems allow for intense factory testing using real equipment interfaced with environment simulators. Almost all functions of the CBTC can and should be tested in the factory environment in order to verify the system and detect most of the anomalies. Integration between the CBTC supplier factory and the field team is key for successful field tests.

2.5.3 Test All Safety-Related Items

All functions related to safety, such as enforcement of speed limit, must be tested in the factory and/or in the field intensively. This statement is also valid for field data such as the position of platforms along the track. Tests of safety functions are used as checks which provide a high-level confidence about the system before revenue service. In addition to completing the safety certification process, CBTC suppliers and authorities having jurisdiction ensure that all safety functions are completely tested before authorizing revenue service.

Functions not related to safety should be tested in order to verify proper operation and minimize impacts on revenue service, but the intensity of tests for nonvital functions should be optimized to avoid unnecessary delays to the revenue service operation.

2.6 Environmental Tests

At the end of the design and before starting the production of the equipment, CBTC equipment is subject to its first set of tests: environmental tests. The different types of environmental tests are as follows:

- ElectroMagnetic Compatibility (EMC) tests
- Climatic condition tests
- Mechanical condition tests
- Abrasive condition tests

Transit agencies and CBTC suppliers establish an environmental test plan to agree on what equipment belongs to each category. The category of the equipment is used to define the test requirements as described in the appropriate test standard. In addition, any specific condition of a particular railroad environment is identified based on field's measurements and tested against.

A common practice in CBTC projects is to accept test results performed by the supplier during their product testing or for another project. This is the only time where test results from other projects may be considered. Difference between the equipment tested on other projects and the equipment for the current project should be analyzed very carefully during this process.

2.6.1 EMC Tests

The goal of the EMC laboratory qualification tests is to verify the level of electromagnetic immunity, that is, susceptibility, and the electromagnetic emissions of the CBTC equipment. These tests are performed in a certified laboratory. As described in [8], which summarizes the EMC activities in CBTC projects, there are two major standards for this activity: (1) CENELEC (European Committee for Electrotechnical Standardization) Standard EN50121 Railway Applications Electromagnetic Compatibility [9] and (2) International Electro-Technical Commission IEC 62236 [10] with the same title. The standard's criteria may need to be adapted to take into account the specificities of CBTC in a railroad environment. For instance, some CBTC radios use the 5.8 GHz frequency so the test should be performed up to 6 GHz instead of the 1 GHz as currently required in the standards.

2.6.2 Climatic Conditions

Climatic condition verifications are related to extreme temperature and humidity tests. Cycles of very cold then ambient temperatures and cycles of very hot then ambient temperatures are tested as per MIL-STD-810F Test Method standard for *Environmental Engineering Considerations and Laboratory Tests* [11]. Also, cycles of dry and humidity environments are tested as per [11].

2.6.3 Mechanical Conditions

Mechanical condition verifications are related to vibration and mechanical shock tests. Standards used are IEC 61373 for rolling stock equipment [12], EN 60068-2-64 on vibration tests [13], and EN 60068-2-27 on shock tests [14] for wayside equipment. It is common that vibration tests help reveal issues with the mechanical design, and therefore among the environmental tests, the vibration test is one of the most important.

2.6.4 Abrasive Conditions

Abrasive conditions are verifying dust and water protections. The size of dust flakes and the amount of water are tested as per IEC 60529 Degree of Protection provided by enclosures [15] depending on the IP code claimed for the equipment. Another type of abrasive testing is the salt fog that can be tested as per [11]. This test is especially important for roadway equipment which is subject to the most severe conditions.

2.7 First Article Configuration Inspection

First Article Configuration Inspection (FACI) is the first check using real equipment by the transit agency that happens in CBTC projects. This activity happens toward the end of the design phase of the project after the equipment has been designed on paper and before starting mass production. The FACI's goal is to verify that production hardware complies with design configuration, drawings, and software design. It provides a means to verify that all documentation is ready for mass production. It also includes the evaluations of maintainability and accessibility. This activity typically takes place at the point of assembly after completion of environmental qualification tests of the prototypes.

2.8 Factory Tests

Verification and validation, as described in the *International Council on Systems Engineering* handbook [16], should be done as much as possible in the factory. CBTC systems allow this type of intensive factory testing and only a few parts cannot be completely tested in the factory. For instance, the interface with the rolling stock cannot be completely tested in the factory. Other parts that cannot be tested are the radio coverage and the database for field data. Functions related to the management of a large number of trains are also difficult to test at the system level. Despite the use of powerful tools such as integrated system factory setup and simulators, it is frequent that anomalies that could have been discovered in the platform are discovered on-site.

2.8.1 Factory Test Goals

The goals of the factory tests are described in [1]. Most important ones are summarized below:

- Test all internal interfaces: All messages including commands and control between ATS and CC, ATS and ZC, and CC and ZC are tested with real equipment. When real equipment is used, the term "tests on target equipment" is employed. However, when the actual software is run on a machine that only emulates real equipment, the term "test on host machine" is used.
- Test every function of the CBTC: Based on the system functional requirement, all functions are tested at a minimum of one location on the line.
- Test external interfaces as much as possible: Interface between the ZC and the wayside conventional signaling system needs to be tested completely. The most numerous interfaces concern the ATS, which needs real or simulated systems to exercise the interface in the factory. The CBTC supplier may develop those simulators based on their knowledge of the external system or the simulator may be provided by the supplier of the external system.
- Perform failure scenarios, especially those which are difficult to perform on-site: for instance, using simulated equipment allows for corrupt messages not possible on-site with the real equipment.
- Support field testing by reproducing issues: The field team's goal is to demonstrate that the system works properly and not necessarily to investigate issues. Once an issue has been detected on-site, it is reported to the factory test team that reproduces and analyzes it with the help of the designers and developers who are available in the same office.

2.8.2 Factory Setup

It is very important that the system factory setup include real hardware for each subsystem. It should include the ATS servers, CCs for checking redundancy, and ZCs for checking redundancy and handover between ZC territories. The factory setup should contain at minimum two real CCs and two real ZCs. Note that during the test, trains are able to run over the complete line, but as there are only a few real ZCs on the factory setup, the ZCs for the zone under test are selected to be the real ZCs, whereas the others are simulated. If the project includes a Solid State Interlocking (SSI) or a programmable logic controller for the interface to the interlocking, then they should also be included. A good practice that is required for large projects is to have a complete factory setup on-site. It allows field teams to run pretests before using track access and also for all personnel involved in the project to become familiar with the system (training, visits, etc). It can be used for regression verifications witnessed by the transit agency for each new version. Though the investment cost is significant, it is a valuable asset during the project and it can be handed over to the transit agency after the project.

2.8.3 Different Types of Factory Tests

2.8.3.1 Product Factory Tests

As discussed earlier, CBTC suppliers use products that are likely applied to several projects at the same time. The product team uses a virtual line that includes all possible track layout configurations. The database used is therefore different from the database for a specific project. Tests at the product level may be very detailed; they test functions that are not part of the system or subsystem specifications known to the transit agency and include failure scenarios that are very low level. The product team is also in charge of verifying that the software runs properly on the target equipment.

2.8.3.2 CBTC Supplier Internal Factory Testing

Factory testing starts with the first version of software even before the design is finalized and the final database is produced. Prior to the factory acceptance test (FAT) witnessed by the transit agency, the CBTC supplier runs all the tests, completes reports that describe all known anomalies, and provides them to the transit agency. Based on the report, the transit agency may decide to hold or to postpone the witness tests.

2.8.3.3 Factory Acceptance Test

The FAT is a major check on the project status and provides a preview of the success rate for the rest of the tests to be performed. It is performed on a software version that is final as much as possible, considering that subsequent field tests will help identify problems, and that a few remaining design issues may still be open. Depending on the project, the FAT is witnessed by the transit agency anywhere from a few days up to a few weeks.

A FAT may be required for each subsystem before a system FAT. However, in most cases, a witnessed subsystem FAT is done only for the DCS.

Because the ATS is the interface to the system operator and it can be tested without a real CC or a real ZC, a separate ATS FAT may also be required by the agency. Indeed, all ATS functions can be tested using real ATS and simulated interfaces. Because most advanced ATS function tests require the complete line and a large number of trains, it may be preferable to only have the real ATS equipment during the ATS FAT.

2.8.3.4 Description of the Tests to Be Performed

First tests to be performed are related to the interfaces. This phase is sometimes referred to as integration testing because it verifies all commands and all controls between the real equipment. This first phase is performed before proceeding to functional tests. This phase may be internal to the CBTC supplier as it does not verify any functional requirement. Then, the core CBTC functions, such as train localization, train tracking, and movement authority enforcement, are tested. Some of the core function tests need to be repeated at all locations. For instance, the train localization should be performed on the entire line in each direction and going over each crossover. Finally, the most advanced and project-specific functions are tested.

2.8.3.5 Major Challenges of Factory Tests

A major concern for the transit agency is that the factory tests be representative of real-life conditions. There are two obstacles to representative tests: (1) the technical aspect of using environment simulators, especially where the DCS is very simplified in the factory setup, and (2) the tests are performed by testers usually not familiar with train operation and therefore may react very differently than the future train operators.

A common issue for CBTC system testing is to assume that the factory setup is responsible for a failure and to frequently reboot all the equipment to have a clean starting point giving more chances for the test to pass. In operation, signaling systems run continuously and cannot be rebooted. To cope with this issue, transit agencies may require endurance tests over several days. Because the number of real CCs and real ZCs is limited, endurance tests are still relatively far from real-life system endurance.

Successful witnessing of FATs is a prerequisite to the start of functional tests on-site. At the end of the tests, some issues will be detected regarding implementation or design. Deciding what issues need to be resolved before the software is sent on-site is a delicate topic. CBTC suppliers tend to minimize the impact of issues, although the transit agency desires that all known errors be corrected. Transit agency concern may be to avoid wasting any track access by having conditions on the test due to a software defect or due to the need of performing the same test again after an issue is corrected. Another transit agency's concern is to maintain a good reputation of the project within the transit agency. If the software first used on-site contains too many errors visible to the train operators during the tests, the agency personnel may be less than enthusiastic with the idea of using CBTC in revenue service in the future.

2.9 On-Board Integration Tests

2.9.1 Rolling Stock Characterization Tests

Though it is not part of the verification and validation, the characteristics of rolling stock should be verified and/or determined through field activities by the CBTC suppliers. The transit agency and rolling stock manufacturer decide based on tests and other considerations what is the guaranteed emergency brake rate and what is the maximum acceleration rate on leveled track. However, it is not sufficient for the CBTC supplier to tune its train model used for proper ATO. CBTC suppliers perform what is called rolling stock characterization tests. The train is controlled by a special CC software. This special software sends commands to the train propulsion and brake system, and then speed sensors on board the train collect data about the train reactions. After the tests, the data are analyzed by CBTC experts in order to determine the train model characteristics. This activity requires several test campaigns where results will be verified during ATO tests with the actual CC software.

2.9.2 Mechanical and Electrical Tests

The first train or first two trains are used as prototypes where the carborne equipment is fitted in the train for the first time. This task includes verifications of the mechanical fit into the car body such as checking alignment of all mounting devices. Once the mechanical interface is verified and issues are solved, electrical test verification is done. Electrical tests start with very low-level tests such as powering the equipment. All hardwire inputs and outputs are exercised using a special CC software. Communication with other electronic systems on board the trains is also verified. This prototype testing usually lasts a few weeks.

2.9.3 Static and Dynamic Post Installation Check Out Tests

Installation tests can be considered CC Post Installation Check Out (PICO) tests where the hardware equipment and the installation are verified. This activity is performed for each and every train using a special CC software dedicated to test along with the first version of the real CC software. In order to verify the speed sensors, a small train movement is involved.

In most projects, all trains are tested with basic CBTC functions, including ATO. Testing ATO on each train is very important because, as noted before, the train characterization test was performed on a limited number of trains. The parameters may have to be fine-tuned to manage the complete train fleet.

To perform verification of CBTC core functions on each train, other equipment, such as radio and ZC, need to be available for at least one location, for instance, on a test track. Similar to the electrical tests described above, all hardwire input, and outputs are exercised using a special software communication to verify that the train was wired properly. Installing and checking CBTC equipment on a train can be performed in a couple of days when there is no problem. A single test shift is typically sufficient for verifying basic CBTC functions.

2.10 Test Track

2.10.1 Use of the Test Track

In brownfield projects, a test track is typically used for testing the CBTC system prior to using mainline tracks. The test track may be used for several purposes:

■ CC tests: To verify the design of the CC/rolling stock interface after the first on-board equipment has been installed.

- System tests: To test most CBTC functions.
- Car by car CC installation verification: To integrate the CC on each train, each train is tested on the test track for core functions in order to verify the interface between CBTC and rolling stock.
- Training: To train the train operators to control the train under CBTC. The test track has to be long enough to serve this purpose.
- System regression tests: To test new CBTC software versions during the deployment phase, as well as after the CBTC has started revenue service.

2.10.2 Test Track Equipment

IEEE recommendations for CBTC testing [1] provide a complete description of the capabilities of an ideal test track. A test track must be equipped with all CBTC equipment: Transponders must be installed on the roadbed, radio access points must be installed near the track, and a network must also be deployed to have communication between the CC and the ZC installed in the technical room. In order to test most CBTC functions, an ATS server and a workstation must also be available, for instance, to set up a slow speed order.

Ideally, the test track layout includes all possible configurations. For instance, all possible types of signals are present. A switch between tracks is also useful. It is common that the test track database includes one or several virtual platforms to test door operation and some of the ATS trip assignment functions. Virtual platforms do not exist in the tracks. They are only in the database for test purposes. The length of the test track is a key element to determine what functions are testable on the test track. To test the ATO operation properly, the train must have enough distance to run at a speed close to the operating speeds on the mainline. Even if the test track does not include all possible configurations, a test track is a must in brownfield projects to help avoid wasting mainline track access at the beginning of the project. In greenfield projects, the test track is the first section of track available for testing.

2.10.3 Location of the Test Track

The test track can be located in the yard area with a specific ability to be configured as mainline in order to test the mainline functions such as ATO. Express tracks can also be used as test tracks during nonpeak hours.

Another option is to use a test track outside the property of the transit agency. With this option, track access is much easier than on the final site, the tests can be performed during day time, and the test track can be long and include switches, grades, and any other configurations. However, travel is required for performing the tests and transportation of rolling stock is necessary. Another issue with remote test track is that the design and installation investment for the test track is not going to be used in the final system. In the driverless migration project presented in [17], a test track connected to the rolling stock manufacturer plant was used. The test track was used by the rolling stock manufacturer for several other projects at the same time.

2.11 On-Site Tests

2.11.1 Post Installation Check Out

Before functional testing can begin, verification of the hardware installation is performed on each and every piece of equipment of the CBTC. For instance, cable and wiring, grounding, and power to the equipment are tested in what is called PICO test. This test is performed for each piece of equipment and constitutes the transition between the installation phase and the testing phase.

Even though this test targets hardware verification, each equipment is booted in order to verify that the operating system on CC, ZC, and ATS works properly and the identification of each equipment is correct. In addition, the network configuration may also be tested. The current software version at the time of the PICO is installed for the CC, ZC, and ATS in preparation of function tests. The software version used during PICO is the factory version at the time of the test; it does not have to be subject to FAT beforehand as no functional test is performed.

This test starts with the first equipment being installed on the property and continues throughout the project. Once a part of the system is in revenue service, care should be taken with PICO because, though very unlikely, adding equipment on the network may affect part of the system in revenue service. To mitigate the risk of service disruption, after one part of the CBTC system is in revenue service, PICO tests are performed during nonrush hours.

2.11.2 DCS Tests

The DCS contains two parts: one for the wayside network between equipment in technical rooms and the control center, and the other for the radio system which may also include a separate wayside network system.

2.11.2.1 Wayside Network Tests

Among the different CBTC subsystems, the first subsystem to be installed and subject to test is the DCS. To perform testing of the DCS, fiber optic between locations must be installed and verified by the installation team. In the technical rooms, network switches are installed and PICO tests performed beforehand. The network management system (NMS) that administrates the network and is provided by the network supplier must also be present before network tests are planned. Tests on the network may include the following:

- Checking of the configuration of each switch
- Verification of connectivity
- Verification of data transfer
- Verification of NMS capability
- Failure scenario of one node to test reconfiguration of the network
- Latency tests
- Throughput tests

During the DCS factory tests on the network, the same tests are performed but on a simplified version of the system with far fewer switches. The factory configuration is supposed to be representative of the on-site network.

Ideally, PICO tests of the network equipment are performed first, then the network is tested before performing PICO of the other CBTC equipment (CC, ZC, ATS) so that communication between equipment can be checked during PICO.

2.11.2.2 Radio Tests

The radio tests are dependent on the type of radio being used on the project. Where the radio system uses a different wayside network than the wayside network used for other communication between technical rooms, the radio wayside network is tested with the same method as explained in the previous paragraph. After the radio wayside network is tested, radio access points are subject to PICO test that includes a power emission test.

Following radio wayside network and access point PICO tests, the radio system tests are performed. First, radio coverage is verified using a train equipped with CBTC. In most CBTC projects, the radio system uses redundant architecture to provide redundant paths for the communication. Redundancy may be both on board the train where there is a radio system at each end of the train and on the wayside where access points may have several frequencies. Where there is frequency diversity, access points of each frequency are installed alternatively along the tracks. A radio tool on board the test train is used to measure the electromagnetic field with the radio link. On some projects, the radio tool can also be used to send, almost continuously, data packets through the network to monitor the number of packet losses. Both ends of the trains may be connected to the tool so that the tests are performed in parallel for each end. In order to verify radio coverage, the first test is done with the access points of only one frequency. Access points for other frequencies are turned off. This test is repeated for each frequency. When performing this coverage test, it is preferable that the train runs at a low speed. This test also verifies that the transition between access points is performed seamlessly. The worst-case configuration for radio coverage may be identified and tested. It corresponds to a specific location where access points can be far apart and there may be one or two masking trains between the train under test and the near access points. Test at maximum operating speeds is also performed. Finally, radio coverage with both train antennas and all access points is verified everywhere.

Testing the radio system is very important to prepare future functional tests of the CBTC system. Where there is a gap of coverage, functional tests other than localization tests are very difficult to perform. Radio tests can be done with a train that is not controlled by the CBTC, meaning that it is possible to test the radio system with a train in revenue service running in the bypass mode in between other revenue trains. Depending on the CBTC product, radio coverage can be performed before or after or during train localization tests. However, in most projects, train localization is tested first to help identify the location of radio coverage gaps.

2.11.3 Localization Tests

Localization tests verify that the train is able to initiate its localization properly, and that the train maintains knowledge on its position on the entire network. Train localization initialization is performed by running over several transponders located on the roadbed, then an odometer system is used to compute the train position based on the distance traveled from the last transponder. The odometer has some uncertainty, typically a small percentage of the distance traveled since the last transponder. To avoid having a large position uncertainty that diminishes CBTC performance, transponders are located on the roadbed at maximum every thousand feet. Each time the train runs over a transponder, the localization error is reset to the minimum. In addition to an equipped train, the prerequisite to start localization tests is to have the transponder installed on the roadbed.

Similar to the radio tests, localization tests can be done with a train in the bypass mode where the CBTC is not controlling the train. Where the radio network is not available or the interlocking is not yet connected to the CBTC, the train may delocalize on diverging switches because positions of the switches are unknown. Train localization is reinitiated soon after traversing a switch, thanks to transponders being placed close to the switch area. Complete localization tests are possible only after the switch positions are provided to the ZC, and the ZC communicates with the CC through the radio.

The localization function and radio test are the first two tests to be performed in the field. They are the foundation of CBTC. Though they represent only a few of the CBTC functions, these tests constitute a large portion of the track access time because the tests are performed on all tracks in both travel directions.

2.11.4 Integration Tests

2.11.4.1 Integration Tests: Internal to CBTC

The term "integration" is used when verifying that new equipment is connected to the system and when checking functional communication between two subsystems. This test is not always witnessed by the transit agency. To perform this test with the CC, radio coverage must have been checked already. Depending on the status of the communication network at the time of the PICO, the internal integration test is performed as part of the PICO or later. For instance, this test verifies that the ATS is able to monitor the status of the equipment, and that commands can be sent to the equipment, basic commands such as active unit switch over. Where track access is very limited, the goal of this test is to prepare for functional testing, which involves more resources such as test personnel and trains.

2.11.4.2 Integration Tests: External to CBTC

For hard-wired external interfaces to the CBTC, all inputs and outputs are tested during this integration phase. The functional part is left aside as much as possible to facilitate testing. For instance, all track circuits and switch positions are verified. External interface tests can be performed independently from the localization and the radio tests. Ideally, they are performed in parallel.

For data communication interface tests such as interface to an SSI, there may be a difference between the transit agency that prefers to test every bit and the CBTC suppliers who want to only sample the bit map. The argument from the CBTC suppliers to perform only sample tests is that each bit was previously tested in the factory. The argument from the transit agency is that the factory tests are not witnessed and the configuration files may have changed since the factory tests. In an ideal world where every change is tracked and retested in factory, there is no need for testing all external digital interfaces. Because external interfaces may involve another supplier, retest after every new version might not happen in the factory and field testing is then necessary. The decision is project dependent based on how much each interface changes and how well the CBTC supplier works with the external system provider.

2.11.5 Functional Tests

After the localization tests, the radio tests, and the interface tests, CBTC functional test can begin. With the exception of the ATO, in an ideal world, the CBTC system should be working properly. Indeed, all functional tests may be performed in the factory, and, therefore, the functional tests need only to be performed to demonstrate the functions to the transit agency for acceptance of the system. However, in the real world, several months of testing and corrections are required. It is common that the transit agency requires that most tests be performed in the field even though they were already performed in the factory. When all functions work properly, performing site acceptance tests can be done within weeks.

The first functional test to be performed is train tracking; it can also be done with a train in the bypass mode not being controlled by the CBTC system. Tracking by the ZC and the ATS is tested at the same time and can be considered a low-level CBTC test. If the communication is ready and the equipment has been subject to PICO, the tracking tests may be performed at the same time as the localization tests.

Following the train tracking, the CBTC basic functions are verified, although at the same time all field device positions are checked. For instance, the verification that the CC does not let a train pass a red interlocking signal is performed for each interlocking signal. Safety-related field items such as interlocking switch position or track circuit boundaries are verified. The test goal is to verify the presence of the field item in the database, and that the proper type of item has been set up in the database. The actual position of the field item in the database is only demonstrated, which is not proven by test. Position of field equipment in the database is validated, thanks to a rigorous process of verification of the surveyed data and database creation. Once all the basic functions are verified, more advanced tests can be performed. Failure scenarios of the CBTC and ATS are performed. Also non-safety-related functional tests are performed.

The art of testing CBTC is in choosing the functions to be tested along with what needs to be repeated at every location in the field. Because CBTC is such a complex system, it is not possible to verify each function in all conditions at every field location. Tests must focus on safety-related functions to build confidence in the system. Factory tests are not sufficient to provide enough confidence to the transit authority to let a system start revenue service without witnessing on-site that the system behaves in a safe manner. Therefore, all field items related to safety such as interlocking areas and all safety functions such as work zone restriction should be tested. The dilemma about how much to test starts with safety functions that are not related to the database, for instance, slow speed order and work zone. There is a near very large number of slow speed order, speed selection, and start and end points. How many slow speed order tests to conduct depends on the design of the system so the CBTC supplier should indicate the type of test to perform. It is common to perform a slow speed order test for each ZC, where only one test should be sufficient because the same software runs on every ZC; only the database changes from one ZC to another.

Regarding the non-safety-related functions, minimum tests are performed to limit track access. For instance, the test to open and close the train door from the ATS when the train is berthed at platform does not need to be done for each platform. However, the safety-related function to check that the CC enables door opening on the proper side of the track and only within the platform boundary is performed for each platform.

Most of the CBTC issues found on-site are not related to any preidentified tests. The tests that verify the CBTC requirements have already been performed in the factory, and, therefore, it is unlikely that a new error that cannot be discovered in factory is found on-site. Special tests that were not identified during development of the factory and field test procedures, because they do not correspond to any requirement, are where most issues are detected in the field. It is important to have official procedures to use during testing, but most issues will not correspond to a specific test step in an official procedure. The majority of the problems are found during the first runs using CBTC and are evident; for instance, the CC does not release the brakes or the ZC does not provide the MAL. After the first months of debugging, the performance of functional tests is more likely to reveal design issues than implementation ones.

2.11.6 ATO Tests

ATO is used in most CBTC projects. ATO tests are performed first on the test track to verify that the ATO software is able to control the train properly such as maintain appropriate speed and make proper station stopping accuracy without being overriden by the ATP function. To test every speed and every change of civil speed, ATO tests should be performed on each track and in both directions. During ATO testing, station stopping accuracy is also verified.

It is preferable to perform the CBTC basic functional tests in the ATO mode rather than performing the test when the train engineer is controlling the train under ATP supervision. Doing so, both the stopping point of the train and the field item locations are checked. As the ATO mode may not be available for testing at the time CBTC basic functions are tested, the test may need to start in a manual mode of operation.

2.11.7 ATS Tests

Except for integration tests when each command and control is verified, most ATS functions require the other CBTC subsystems to work properly. Some main advanced ATS functions include automatic train routing used for setting up routes and managing train junctions, and trip assignment based on the train operating schedule. For driverless projects, the ATS is needed to perform CBTC functional tests and the project schedule must be attentive to have all CC, ZC, and ATS from the factory ready for field testing at the same time. Because all ATS functions, except route setting and interface testing discussed before, can be tested very well in the factory, specific tests of the ATS can be limited on-site. The advanced ATS functions such as train regulation are part of CBTC system tests where the CC and the ZC are also tested at the same time. The route setting function is particular because a specific check must be done for each interlocking signal on-site to confirm that the routes are set early enough so that the test train does not have to slow down in front of a red signal.

2.11.8 Site Acceptance Tests

Depending on the project, the transit agency witnesses and approves results of only the functional and external interface tests, or all tests, including the PICO and both internal and external integration tests. The site acceptance tests are written based on the system functional specifications derived from the contract. All requirements that can be tested on-site are tested, and additional tests identified by the CBTC suppliers should also be performed. To meet the revenue service target date, it is common practice for the first version of the CBTC software to not include all functions. Advanced functions, often related to the ATS, are pushed to a later stage. Any nonurgent issue detected on the first version is corrected in the version that includes the advanced functions.

Prior to the site acceptance tests, the transit agency may want to witness a sample of the debugging phase of the CBTC tests in order to stay informed on the progress and to increase the transit agency knowledge of the system. This corresponds to the phase where test procedures are checked before being performed officially and when issues are investigated and solved by the suppliers.

2.11.9 Shadow Mode Tests

A concept used in CBTC migration projects is to monitor the performance of the CBTC system for an extended period of time, although it is not yet controlling the trains or any field devices. This period is called "shadow mode," "ghost mode," or "monitor mode." All inputs are active and connected to the CBTC equipment, but the CBTC CC is not controlling the train and the interlocking system is not applying any commands from the ZC. This type of test starts prior to starting revenue service and typically lasts several months. This phase helps demonstrate the reliability and availability of the CBTC system, and it helps build confidence among the stakeholders for the project. When certification for revenue service is approved, trains controlled by the CBTC without passengers are introduced in between revenue service trains. Once the shadow mode period is completed successfully, passenger revenue service can begin for the CBTC system provided the safety certificated was obtained previously.

2.11.10 Reliability, Availability, and Maintenance Tests

2.11.10.1 Reliability and Availability Demonstration

The goal of the reliability and availability demonstration is to demonstrate, toward the end of the project, that the characteristics of the system actually meet the contract requirements regarding reliability and availability, measured in terms of mean time between failures (MTBF), mean time between functional failures (MTBFF), or other measurements defined in the contract.

In CBTC projects, the demonstration generally lasts at least 6 months and begins immediately after revenue service has started on the entire line. All equipment need to be installed and turned on continuously for the test to be representative. Six-month duration is the minimum length in order to measure the reliability with a good level of confidence. The success of the test may be a condition to exit the warranty period, which is typically 2 years after the last CBTC equipment has entered revenue service.

The availability is calculated using the total test time minus the observed downtime due to CBTC failures divided by the total test time. Definitions of MTBF and MTBFF are the number of failures and the number of functional failures over the accumulated operating time for one type of equipment. The definitions actually apply to the entire life of the system and would not be representative of the system in a 6-month test. Therefore, it is common that CBTC contract requires a specific criteria based on a one-sided chi-square test with 90% or 95% of confidence level. This test has several other names such as goodness-of-fit tests. Both the CBTC supplier and the transit agency should agree on the terms and conditions of the test when signing the contract or before starting the demonstration.

2.11.10.2 *Maintainability Demonstration*

Maintainability is the ease with which a product can be maintained. Maintainability demonstration is performed after the transit agency maintenance personnel have been trained. It can be performed before or after the system has started revenue service; for instance, it can be done during the reliability and availability demonstration. The maintainability demonstration serves two purposes: verify that the mean time to repair (MTTR) meets the contract requirement and show that the training and maintenance manuals are adequate.

In order to verify the MTTR, the equipment to be tested has to be agreed by the CBTC provider and the transit agency. One approach is to use a statistical method for sampling the population of CBTC items. Producer and consumer risks are used to determine a sample size. Once the number of items to be tested is calculated, the agency and the supplier choose randomly the equipment to be tested. Another mathematical method of choosing the tests to be done is explained in [18]. This method is based on the failure rates and the number of identical items in the system. In this method, the agency can decide on testing the items that will represent a high percentage of the maintenance actions.

When the test is performed, a failure is introduced on purpose and the repair is done by the transit agency maintenance personnel without the assistance of the CBTC supplier team. It is a good practice to have several teams of maintenance personnel doing the same corrective action and use the mean repair time for the same correction. The results are gathered in a report after the tests, and an MTTR is calculated based on the mean of the repair duration observed on-site. One of the possible positive outcomes can be updated maintenance procedures.

As suggested in [4], MTTR is usually required for the complete CBTC system. An issue with the mathematical approach proposed in [18] is that as the number of CCs is much larger than the other subsystems, most of the maintenance will thus be on the CC and the mathematical method results in choosing only CC equipment. The analysis should be performed subsystem by subsystem to avoid this issue.

2.12 CBTC Test Duration

The duration of the field CBTC testing phase depends on various factors, and it is not possible to define test duration without knowing project specifics. However, based on the examples from projects in the last decade in different countries by different suppliers, a range of duration is provided in this section.

As discussed previously, CBTC projects are of two types: greenfield projects on new lines and brownfield projects on existing lines where revenue service must be maintained during the project. For greenfield projects, the total duration of the project can be planned in as little as 2 years, whereas the minimum time for a brownfield project is about 5 years. CBTC projects are often more difficult than anticipated and important delays are very common, especially with brownfield projects where a 5-year project is considered successful when executed in less than 7 years.

The experienced delay between the first CBTC field tests and the revenue service is at least 1 year. For brownfield projects where CBTC starts revenue service on a section by section basis, the testing phase for the full line may last up to several years.

In order to have an earlier revenue service date, a common practice is to postpone the inclusion of advanced functions in the first CBTC versions used for revenue service. The advanced functions and subsequent changes are tested after the CBTC starts revenue service. Therefore, in addition to fixing issues discovered during first months of revenue service, the supplier performs tests for these advanced functions. It is worth noting that in many projects, CBTC testing is not completely over when CBTC starts revenue service. Converging to the final error-free service may take as much time as original field testing.

2.13 Constraints on Field Tests

Testing CBTC is a difficult task for several reasons. The most common issues encountered are as follows:

- Difficulty to obtain track access: This is especially true on brownfield projects where the line is already in revenue service. The tracks are used during the day for revenue service operations and during the night or off-peak times for maintenance. For greenfield projects, track access is relatively easier to obtain, but still it is competitive because all other subsystems may also need the tracks to complete their installation and tests.
- Coordination with the installation: During integration tests, where field elements are verified one by one, a prerequisite is of course the equipment installation. The long-term planning, including the equipment installation order, must be reviewed carefully to match with the test planning. Because CBTC projects are very large with many stakeholders and companies involved, it may happen that testing is planned and testers discover that the equipment

to be tested is not installed. Communication between installation and test teams is very important and usually requires some adjustments at the beginning of the project.

■ SSI interface: CBTC can be used as a standalone system, or it can be overlayed over conventional signaling system. When the CBTC system is overlayed over a conventional signaling system, the system is considerably more complex and requires more testing but the major impact is the potential delay of the interlocking deployment. Renewing a relay-based interlocking with an SSI is a challenge. Adding to this challenge, the new interlocking functions that manage the CBTC system make the SSI project very difficult and subject to potential delays. CBTC functions need the interlocking to be tested completely, and, therefore, any delay in the interlocking project may directly affect the ability to test the CBTC system.

Other common issues that may impact the testing are as follows:

■ FAT: It is required that the FAT be successful before tests can be performed on-site. Depending on when the design is finalized and to a lesser extent depending on the efforts on factory tests by the CBTC supplier, the FAT may be delayed which has the potential to delay the beginning of field tests. Note that a delay in finalizing design might also affect installation.

■ Support from factory: An important factor in the success of field testing is the ability for the field team to obtain support from the factory test team and the designers. During the field testing phase, the field team identifies and reports issues to the factory team. The factory test team then analyzes the issues and fixes them in a new version; help from the designers may be necessary. New versions of software and patches are frequent and required to converge to error-free software. The time delay between detection of an issue and the next software version containing the fix are critical and dependent on CBTC supplier factory teams. Projects that are linked to special events such as Olympic Games cannot be delayed and therefore are priorities for the company. While a priority project is being taken care of, other projects might suffer from a lack of resources.

2.14 Conclusion

CBTC technology is used by transit agencies all over the world mainly for its ability to minimize headway between trains and for safety through continuous speed control. It is used in most new urban transit lines and also for renewing signaling systems of existing lines. As for every large and complex engineering system, testing of a CBTC system is very challenging. Fortunately, CBTC systems are able to be tested almost entirely in the factory. Despite the factory testing, many field tests need to be performed, some specific to the field and some as a repeat of the factory tests. Repeating some factory testing is technically not necessary, but this is how the industry has been working so far. Specific field tests are common to all CBTC projects to verify core functions and field data. Choosing what tests to perform in the factory and on-site is based entirely on the experience of the CBTC supplier and the transit agency. Despite intensive factory and field testing, even on projects with a very long field test phase, it is inevitable that the CBTC will experience new conditions during the first months of operations, which might lead to an operational disturbance. Hopefully, in the future, CBTC suppliers will improve their system development and their factory test capabilities to a level which will be sufficient to effectively minimize field tests. Another option to facilitate the entire signaling project, including the testing part, is to simplify the system and to minimize the customized functions.

References

1. IEEE. IEEE Std 1474.4-2011: IEEE recommended practice for functional testing of a communications-based train control (CBTC) system. IEEE, New York, 2011.
2. A. Rumsey. Implementing CBTC on an operating transit system. *American Public Transportation Association Rail Conference*, Montreal, Quebec, Canada, June 2014.
3. K. Diemunsch. Migration to communication based train control. *American Public Transportation Association Rail Conference*, Montreal, Quebec, Canada, June 2014.
4. IEEE. IEEE Std 1474.1: IEEE standard for communications-based train control (CBTC) performance and functional requirements. IEEE. New York, 2004.
5. IEEE. IEEE Std 1474.2: IEEE standard for operator interface requirements in communications-based train control (CBTC) systems. IEEE. New York, 2003.
6. IEEE. IEEE Std 1474.3: IEEE recommended practice for communications-based train control (CBTC) system design and functional allocations. IEEE. New York, 2008.
7. IEEE. IEEE Std 802.11: IEEE wireless local area network std 802.11 series. IEEE. New York, 2012.
8. N. Glennie. EMC laboratory qualification testing for CBTC equipment. *American Public Transportation Association Rail Conference*. Montreal, Quebec, Canada, June 15–18, 2014.
9. European Committee for Electrotechnical Standardization. EN 50121: Railway applications—Electromagnetic compatibility. European Committee for Electrotechnical Standardization, Brussels, Belgium, 2006.
10. European Committee for Electrotechnical Standardization. EN 62236-1: Railway applications—Electromagnetic compatibility. European Committee for Electrotechnical Standardization, Brussels, Belgium, 2008.
11. Military Standard. MIL-STD-810G: Department of Defense—*Test Method Standard for Environmental Engineering Considerations and Laboratory Tests*, U.S. Department of Defense, Washington, DC, 2008.
12. IEC 61373: Rolling stock equipment—Shock and vibration tests. International Electrotechnical Commission, Geneva, Switzerland, 2010.

13. European Committee for Electrotechnical Standardization. EN 60068-2-64: Environmental testing—Part 2-64: Tests—Test Fh: Vibration, broadband random and guidance. European Committee for Electrotechnical Standardization, Brussels, Belgium, 2008.

14. European Committee for Electrotechnical Standardization. EN 60068-2-27: Environmental testing—Part 2-27: Tests—Test Ea and guidance: Shock. European Committee for Electrotechnical Standardization, Brussels, Belgium, 2008.

15. IEC 60529: Degree of protection provided by enclosures (IP code). International Electrotechnical Commission, Geneva, Switzerland, 2010.

16. International Council on Systems Engineering. *International Council on Systems Engineering (INCOSE) Handbook*, version 3.2.2. International Council on Systems Engineering, San Diego, CA, 2011.

17. H. Murray. Line 1 conversion makes metro history. *Railway Gazette International*, 168(2), 52, 2012.

18. Military Standard. MIL-STD-471A: *Military Standardization Handbook 471 (MIL-HDBK-471)—Maintainability Verification/Demonstration/Evaluation*. U.S. Department of Defense, Washington, DC, March 27, 1973.

Chapter 3

Safe Rail Transport via Nondestructive Testing Inspection of Rails and Communications-Based Train Control Systems

Vassilios Kappatos, Tat-Hean Gan,
and Dimitris Stamatelos

Contents

3.1 Introduction 44
3.2 Overview of CBTC's Capacity 45
3.3 Rail and Fastening Parts Infrastructure 46
    3.3.1 Superstructure Subsection 46
    3.3.2 Rail Overview 46
        3.3.2.1 Defects in Rails 47
    3.3.3 Fastening Parts Overview 49
    3.3.4 Critical Place on Rail Network 49
3.4 Rail and Fastening Parts Inspection 50
    3.4.1 Manual and Automated Visual Rail Inspection 51
    3.4.2 Liquid Penetrant Rail Inspection 52
    3.4.3 Ultrasonic Rail Inspection 52
        3.4.3.1 Conventional Ultrasonic Rail Inspection 53

3.4.3.2 Laser Ultrasonic Rail Inspection...53
3.4.3.3 Phased Array Ultrasonic Rail Inspection 54
3.4.3.4 Rail Inspection Using Long-Range Ultrasonics
(Guided Waves) 54
3.4.4 Magnetic Flux Leakage Rail Inspection..55
3.4.5 Eddy Current Rail Inspection..56
3.4.6 Alternating Current Field Measurement
Rail Inspection 57
3.4.7 Rail Inspection Using Electromagnetic Transducers....................57
3.5 Comparison of NDT Techniques...57
References 62

3.1 Introduction

Trains constitute one of the most popular and efficient means of passenger and freight transport all over the world. The rapid and continuous increase in train traffic, train speed, and tonnage carried on the rail network has posed the challenge of ensuring that rail travel remains a safe, attractive, and on-time mode of transport for people and goods to public authorities and railway companies.

Failures and unavailability of railway infrastructures* and rail operations can have catastrophic consequences. One of the most common precursors to catastrophic accidents on the railway is signal passed at danger† (SPAD) [1]. A high number of SPAD incidents happen each year, even though the majority of them are highly unlikely to result in a serious accident, due to the high probability of instant braking and/or the low speeds observed. However, a safety violation in Chatsworth, California, in September 2008, resulted in a collision between a freight train and a commuter train [2], causing the death of 26 people and more than 135 people injured. Another accident occurred in Macadona, Texas, in June 2004, when a train passed a stop signal without authority to do so, which resulted in three deaths and 30 people injured [3].

Apart from the SPAD accidents, many accidents were caused by the catastrophic failure of rail components, such as in Hatfield, Hertfordshire, in October 2000, that led to the loss of 4 lives and 70 people injured [4] and in Minot, North Dakota, in January 2002, when a freight train derailed, spilling gas and hazardous materials, killing at least one person, and injuring around a hundred more. The National Transportation Safety Board determines that the probable cause of the

---

* The term "railway infrastructure" covers all assets used for train operations, except rolling stocks. A definition of railway infrastructure is given by the European Community Regulation 2598/1970 and comprises routes, tracks, and field installations necessary for the safe circulation of trains.
† A SPAD occurs when a train passes a stop signal without authority to do so.

train derailment was the misidentification and nonreplacement of the cracked joint bars [5].

Almost all the accidents and the associated casualties caused by SPAD and failure of rail components could have been at least minimized or altogether prevented, as long as train protection (TP) equipment had been installed and operated, and through careful nondestructive testing (NDT) inspection and appropriately scheduled maintenance of rails, respectively. Both TP and NDT of rails are significant, safety-critical applications, which show very demanding requirements in terms of availability, continuity, and integrity. In order to fulfill these high-performance demands, it is essential that both terms are understood fully.

This chapter mainly studies the NDT techniques that can be employed to inspect rails and fastening parts as well as relevant research and development work in this field. As the NDT techniques significantly depend on the nature of defects, a discussion about the defects that emerge on the rail infrastructure takes place. Finally, in Section 3.2, an overview of the capacity of the recent TP methods mainly based on the communications-based train control (CBTC) is carried out for a complete overview of all measures (NDT, TP) that can be used to avoid any potential and serious rail accidents.

3.2 Overview of CBTC's Capacity

Over the long term, it was easier to develop devices to protect against signal errors than driver errors, but by the late 1980s electronics had developed to the point at which it was possible to protect against driver errors by installing systems that continuously supervise the movement of trains and automatically apply the brakes if a train is going too fast for current track and signaling conditions. TP is equipment fitted to trains and the track that can reduce risks from SPADs and overspeeding. There are many different ways of preventing SPADs or reducing their effects, including different types of automatic TP (ATP) in the United Kingdom and the European Union, and positive train control (PTC) in the United States. Generally, ATP refers to either of two implementations of a TP system installed in some trains in order to help prevent collisions through a driver's failure to observe a signal or speed restriction.

Railway signaling systems are essentially used to prevent trains from colliding. One of the railway signaling systems that makes use of the telecommunications between the train and the track equipment for the traffic management and infrastructure control is the CBTC[*]. CBTC has been under development since the mid-1980s [6–8]; however, wide-scale adoption has not occurred because of

---

[*] As defined in the IEEE 1474 standard (IEEE, 1999), a CBTC system is a "continuous, automatic train control system utilizing high-resolution train location determination, independent of track circuits; continuous, high-capacity, bidirectional train-to-wayside data communications; and trainborne and wayside processors capable of implementing ATP functions, as well as optional automatic train operation (ATO) and automatic train supervision (ATS) functions."

technical, practical, economic, and institutional barriers [9]. Recent regulations and legislation have altered the situation [10] and have mandated PTC's implementation on a large portion of the main lines by 2015.

A number of studies have previously investigated the impact of CBTC on the capacity. Lee et al. determined that moving blocks could increase the capacity of the Korean high-speed railway [11]. Another study quantified the capacity benefits of the European Train Control System, Europe's version of CBTC [12]. In the United States, Smith et al. studied the potential benefits of Burlington Northern's Advanced Railroad Electronics System and other possible CBTC systems [13–16]. They calculated how the more efficient meet–pass planning and the increased dispatching effectiveness possible with CBTC will affect the capacity. Martland and Smith calculated the potential terminal efficiency improvements resulting from the estimated increases in reliability offered by CBTC [17]. Many authors have claimed that a CBTC system with moving blocks will increase capacity [7,13–15,18]. CBTC makes the train, signal, and traffic control systems more "intelligent" [19], allowing the railroad to better plan and control train movements, increasing railroad efficiency and capacity.

3.3 Rail and Fastening Parts Infrastructure

3.3.1 Superstructure Subsection

Maintenance of rail infrastructure can refer to the following components: maintenance of *track* (structured into superstructure and subgrade subsections), *bridges and tunnels, electrification equipment, signaling and communication equipment, rail traffic*, and so on. The superstructure is subject to periodical maintenance and replacement. A survey conducted among maintenance agencies [20] revealed that most maintenance activities of superstructure subsection are concentrated in rail and all fastening parts maintenance.

The elements to be considered in the superstructure (Figure 3.1) are mainly *rail*, which supports and guides the train wheels; *sleepers and all fastening parts*, which distribute the loads applied to the rails and fix rails to railroad ties; *ballast*, which usually consists of crushed stone—and gravel only in exceptional cases—and should ensure the damping of most of the train vibrations, adequate load distribution, and fast drainage of rainwater; and *sub-ballast*, which consists of gravel and sand. This protects the upper layer of the subgrade from the penetration of ballast stones, whereas at the same time it contributes to distributing further external loads and ensuring the quick drainage of rainwater.

3.3.2 Rail Overview

Rails are longitudinal steel members that accommodate wheel loads and distribute these loads over the sleepers or supports, guiding the train wheels evenly and

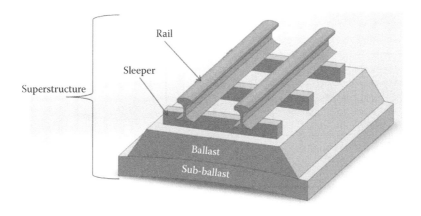

Figure 3.1 Superstructure subsection.

continuously [21]. There is a range of rail cross sections, materials, weights, and head profiles used worldwide. Mainly, rails are made from high steel, which has led to a significant improvement in rail fatigue performance and a considerable reduction in residual stress development [22]. Many standards are used for rail profiles, which are classified into the International Union of Railways, the American Society of Civil Engineers, the American Railway Engineering Association, and the British Standards. Other profiles are used in the Netherlands, Denmark, Germany, India, China, South Africa (SAR), and so on. In Europe, the maximum static axle load ranges about 21–25 ts; in the United States, it normally reaches almost 30 ts and in Australia about 37 ts has been reported on iron-ore vehicles. All these axle loads are nominal values, assuming that vehicles are uniformly loaded [23].

3.3.2.1 Defects in Rails

In service, rails can suffer from two types of defects: *noncritical* and *critical*. Noncritical defects do not affect the structural integrity of the rail and/or the safety of the trains operating over the defect; however, critical defects affect these. Typically, the parts of a rail where defects can usually be found are head, web, foot, switchblades, welds, and bolt holes. Although the majority of defects are located in the head, they can also be found in the web and foot. Figure 3.2 demonstrates the propagation of a single crack in the rail head.

Rail defects have been classified in many ways. Depending on their initiation, defects can be divided into three wide groups [24] (Figure 3.3):

1. *Cracks caused by manufacturing defects* [25]. Rail manufacturing defects are generally a result of nonmetallic origin or wrong local mixings of the rail steel components that, under operative loads, generate local concentration of stresses, which trigger the rail failure process [26].

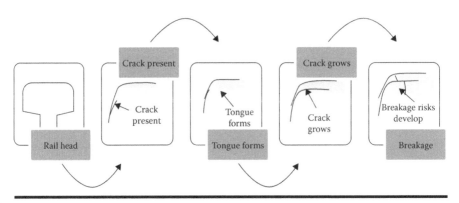

Figure 3.2 Typical crack propagation in the head area of rail.

Figure 3.3 (a) Broken rail (Data from Wikipedia, http://en.wikipedia.org/wiki/ Rail_inspection.), (b) examples of severe loss of rail foot due to severe corrosion (Data from Network Rail.), and (c) shelling.

2. *Defects due to inappropriate handling and use.* These kinds of defects are generally due to spinning of train wheels on rails or sudden train brakes, that is, the wheelburn defect.
3. *Rail wear and fatigue defects* where three defects are the most frequent: (1) *corrugation*, (2) *bolt hole cracks*, and (3) *rolling contact fatigue (RCF).* Corrugation is an event strictly correlated to the wearing of the railhead, generated by a wavelength-fixing mechanism related to the train speed, the distance between the sleepers [27], friction [28], and so on. Bolt hole cracks account for about 50% of the rail defects in joined tracks [29]. RCF damages initiate on or very close to the rail rolling surface [30] and are much more severe in terms of the structure integrity [24,30–31].

3.3.3 Fastening Parts Overview

Fastenings are classified into *rigid* and *elastic*. Rigid fastenings are used only with timber or steel sleepers where the rail is connected to the sleeper with bolts or nails (Figure 3.4a). Elastic fastening systems are widely used and usually constitute toe insulator, side post insulator, rail pad, shoulder, positive lock-in, and clip (Figure 3.4b). The toe insulators are installed into the parked position, where they mechanically hold the side post insulators and pads in place. The clips clamp the rail onto the rail pads with a spring force per clip of up to 1.25 tons. Therefore, the total holding force to rail from each sleeper can be up to 2.5 tons.

The defects in fastening parts can be mainly classified into (1) broken bolts or unhammer nails, (2) gaps between nails and rail, and (3) loosening of a fastening element. Defects in rail and fastening parts are very severe and, if they are undetected and/or untreated, can lead to rail breaks and derailments. There are, therefore, potential hazards for railway safety.

3.3.4 Critical Place on Rail Network

The critical areas in the rail network, where problems are known to occur and the probability of defect initiation and propagation is very high, are usually tunnels, level crossings, bridges, curves, switches, and crossings [33]. The environmental conditions in the tunnels, such as high level of humidity, cause a rapid growth of corrosion and other defects. Level crossings are susceptible to corrosion as well, due to road salt being applied in the areas around the crossing and being further spread in the tire tread of road vehicles. Access and inspection of tunnels and level crossing are very difficult, in particular in the case of level crossings, where the level crossing panels should be removed. With regard to bridges, in Europe there are over 220,000 rail bridges, and most of them are more than 100 years old [34], whereas there are more than 1290 railway tunnels longer than 1000 km [35]. Switches and crossings are subject to more intense stresses than linear tracks. Hence, although the types of defects and required maintenance tasks are similar to those in linear tracks, the behavior and evolution of switches and crossings defects are different than those

(a)

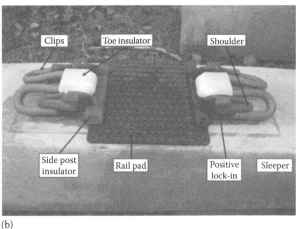

(b)

Figure 3.4 **(a) Common rigid (Data from Wikipedia, http://en.wikipedia.org/wiki/Rail_fastening_system.) and (b) elastic fastening. (Data from Pandrol, The future of rail fastenings, www.pandrol.com.)**

in linear tracks, because of the harder stresses. All these parts are considered very critical because, if they break, this failure makes the train derail. Therefore, the inspection of all these critical places is imperative.

3.4 Rail and Fastening Parts Inspection

The detection of damages in rails and fastening parts at the earliest possible stage is of paramount importance to the rail transport industry. Catastrophic rail failure due to growth of structural defects can be avoided by careful inspection and appropriately scheduled maintenance. The first rail inspections have been carried out manually by human operators, searching for visual surface anomalies. Nowadays, the manual inspections remain in use by means of sophisticated handheld equipment. However, the manual inspection is very slow and laborious, and the results depend on the skill level of the inspector. Relatively recently, rails and fastening parts have been systematically inspected for internal and surface defects using various NDT techniques, such as visual inspection, magnetic induction, eddy currents, photothermal, and ultrasonics [36–40].

By definition, NDT techniques are the means by which materials and structures may be inspected without disruption or impairment of serviceability. NDT is a branch of the engineering science that is concerned with all aspects of the uniformity, quality, and serviceability of materials and structures. The science of NDT incorporates many applied engineering technologies for the detection, location, and characterization of discontinuities, in items ranging from research specimens to finished hardware and products. NDT has become an increasingly vital factor in the effective conduct of research, development, design, and manufacturing programs. A review of current major NDT techniques applicable to railway infrastructure inspection is presented in Sections 3.4.1 through 3.4.7.

3.4.1 Manual and Automated Visual Rail Inspection

Visual rail inspection examines only surface damage, and it can be conducted either manually or using automated visual systems. It is the most common and oldest technique used for rail and fastening parts inspection and is carried out by trained inspection personnel, who walk along the railways and visually check these for any visible defect or damage (Figure 3.5). This technique is neither reliable nor efficient because the chances of human error by missing visible defects are high

Figure 3.5 Manual rail inspections expose maintenance personnel to dangers from passing trains, flying ballast, and projectiles.

and the whole length of the rail needs to be inspected *in situ*, which is very time consuming. Additionally, it exposes the personnel walking along the track to carry out the inspection to danger from moving trains.

To eliminate or reduce, as much as possible, the human error and increase the inspection speed, automated visual systems have been used. The concept of automated visual systems is simple, and it is mainly based on the use of a high-speed camera, capable of capturing video images of the rail as the train moves over it. Afterward, the acquired images are analyzed automatically. Automated visual inspection systems are mostly used to measure surface damage and percentage of wear, pincers position or missing bolts, rail gap, moving sleepers, absence of ballast, base plate condition in absence of ballast, and so on. The operating speed of these systems can vary from 60 to 320 km/h, depending on the type of inspection carried out and resolution required.

3.4.2 Liquid Penetrant Rail Inspection

Liquid penetrant is a visual technique based on the use of special dyes that are spread over the area of interest for inspection, usually a weld. The dye is applied on the cleaned surface of the component to be inspected and allowed to dwell for a few minutes. Once the dye has been allowed to dwell for a sufficient amount of time, the excess dye is wiped away and the developer is applied. If there is a surface defect present, such as a crack or small pits, then the dye that has leaked inside will flow back out after the developer has been applied providing a clear visible indication of the defect. Liquid penetrant is a time-consuming process carried out manually by certified inspection personnel. The technique is extremely sensitive to very small defects only a few millimeters in length and depth but requires thorough cleaning of the surface to be inspected before it can be used [41]. In the case of railways, the technique can be used to inspect welds of rails. The inspection is relatively fast but, due to the large number of components to be inspected, considerable time is required. Only surface-breaking defects are detectable with this technique.

3.4.3 Ultrasonic Rail Inspection

The current existing ultrasonic rail defect detection techniques and systems have limitations in terms of the inspection/monitoring reliability over the entire rail section, mainly concentrated on the rail head. The ultrasonic testing is applied using the ultrasonic vehicles and the walking stick probes only cover defects that are located in the head area, the web, and the part of the foot directly beneath the rail web (Figure 3.6). This is particularly important because a high percentage of rail breaks are caused by defects that initiate in the foot of the rail (Figure 3.6c) and remain undetected due to their location and small size.

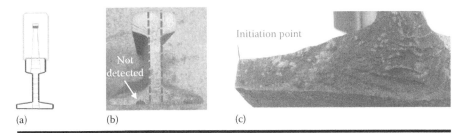

Figure 3.6 Coverage of train-mounted and walking stick probes (a), geometrical limitations on the current ultrasonic inspection (b), and example of a rail break due to small rail foot defect (c). (Data from http://www.networkrail.co.uk/.)

3.4.3.1 Conventional Ultrasonic Rail Inspection

During the inspection of rails using conventional ultrasonic probes, a beam of ultrasonic energy is transmitted into the rail. The reflected or scattered energy of the transmitted beam is then detected using a collection of transducers. The amplitude of any reflections together with when they occur in time can provide valuable information about the integrity of the rail. Ultrasonic inspection is carried out by a variety of different instruments ranging from handheld devices, through dual-purpose road/track vehicles to test fixtures that are towed or carried by dedicated rail cars. A couple of difficulties encountered by the ultrasonic methods are: very cold weather, where ice interferes with testing by providing an intervening interface; damaged rail where a sliver of rail can slice/puncture a tire; heavily applied lubrication can affect the inspection results, producing an intervening interface.

3.4.3.2 Laser Ultrasonic Rail Inspection

Laser ultrasonic testing combines the sensitivity of ultrasonic inspection with the flexibility of optical systems in dealing with complex inspection problems. Its remote nature allows the rapid inspection of curved surfaces on fixed or moving parts. It can measure parts in hostile environments or at temperatures well above those that can be tolerated using existing techniques. Its accuracy and flexibility have made it an attractive new option in the NDT rail infrastructure field.

Laser-based ultrasonics is a remote implementation of conventional ultrasonic inspection systems that normally use contact transducers, squirter transducers, or immersion systems. Laser ultrasonic systems operate by first generating ultrasound in a sample using a pulsed laser. When the laser pulse strikes the sample, ultrasonic waves are generated through a thermoelastic process or by ablation. The full complement of waves (compressional, shear, surface, and plate) can be generated with lasers. When this ultrasonic wave reaches the surface of the sample, the resulting surface displacement is measured with the laser ultrasonic receiver based on an adaptive interferometer. In Reference [42], preliminary tests showed that the developed laser ultrasonic system can be used to inspect the entire rail section including rail head, web, and base.

3.4.3.3 Phased Array Ultrasonic Rail Inspection

The main advantage of using arrays in NDT over conventional single-element transducers is the ability to perform multiple inspections without the need for reconfiguration and also the potential for improved sensitivity and coverage. A particular array is able to undertake a range of different inspections from a single location and so is more flexible than a single-element transducer. In fact, an array can generate ultrasonic fields of almost infinite variety. However, they are most commonly used to produce fields similar to those from traditional single-element transducers, that is, plane, focused, and steered beams. Additionally, most types of arrays (with the exception of annular arrays) can be used to produce images at each test location. This allows rapid visualization of the internal structure of a component.

Electronic scanning permits very rapid coverage of the components, typically an order of magnitude faster than a single transducer mechanical system. Beam steering (usually called sectorial or azimuthal scanning) can be used for mapping components at appropriate angles to optimize the probability of detection of discontinuities. Sectorial scanning is also useful when only a minimal footprint is possible. Electronic focusing permits the optimization of the beam shape and size at the expected discontinuity location, as well as the optimization of the probability of detection. Overall, the use of phased arrays permits optimizing discontinuity detection while minimizing testing time. No practical systems involving ultrasonic phased arrays have been developed for high-speed rail inspection so far, due to the problems that arise from the large amount of data that need to be analyzed.

3.4.3.4 Rail Inspection Using Long-Range Ultrasonics (Guided Waves)

Guided wave testing method is one of the latest developed methods in the field of ultrasonic NDT. The guided wave method usually employs low-frequency waves, compared to those used in conventional ultrasound testing, typically between 10 and 100 kHz. Higher frequencies can be used in some cases, but a detection range is significantly reduced. The method employs mechanical stress waves that propagate along a structure while guided by its boundaries. This allows waves to travel a long distance with little loss in energy compared with unguided waves of the same frequency, and as such numerous and widespread applications of guided wave technology can be found in literature [43–44]. Rails are natural excellent wave guides due to their installation as continuous welded lengths.

Figure 3.7 **Transducer mounting for rail foot inspection.**

In this light, guided wave variations have been developed for rail track inspections [37,45] where typically three sensor setups are encountered: (1) fixed sensors on rail (Figure 3.7), (2) guided wave rail inspection vehicle, and (3) sensor-on-train system [45]. In the first case, the sensors are permanently mounted on a rail, inspecting mainly the areas where the probability for defect detection is very high and the access to carry out the conventional NDT techniques is limited, such as level crossings, switches and crossings (cast crossings), and tunnels. In the second case, ultrasonic transducers are mounted on both ends of the inspection vehicle, whereby energy can be induced into the rail at one end and received at the other end. In the last case, the sensors would be mounted on the train and the ultrasonic energy and vibrational patterns would be propagated forward from the moving train and reflected back with a modified pattern recorded by the transducer if defects were encountered.

3.4.4 Magnetic Flux Leakage Rail Inspection

Magnetic flux leakage is a magnetic method of NDT that is mainly used to detect corrosion and pitting in steel structures, most commonly pipelines and storage tanks. The basic principle is that a powerful magnet is used to magnetize the steel. At areas where there is corrosion or missing metal, the magnetic field "leaks" from the steel. In a magnetic flux leakage tool, a magnetic detector is placed between the poles of the magnet to detect the leakage field. Analysts interpret the chart recording of the leakage field to identify damaged areas and hopefully to estimate the depth of metal loss. In rail inspection using magnetic flux leakage, search coils fixed at a constant distance from the rail are used to detect any changes in the magnetic field that is generated by a direct current electromagnet around the rail. In the areas where a near-surface or surface transverse defect is present in the rail, ferromagnetic steel will not support magnetic flux, and some of the flux is forced out of the part. The sensing coil detects a change in the magnetic field and the defect indication is recorded [46]. The magnetic flux leakage can detect mainly transverse fissures because the flaws run parallel to the magnetic flux lines or the flaws are too far away from the sensing coils to detect.

3.4.5 Eddy Current Rail Inspection

Eddy current inspection techniques have originated from Michael Faraday's discovery of electromagnetic induction in 1831. The principle of eddy current is based on the phenomenon that occurs when an alternating current flows within a coil, causing a changing magnetic field to be produced. If the excitation coils producing the changing magnetic field are brought near the surface of a conductor, regardless of whether it is ferromagnetic or paramagnetic, it will cause electric currents or eddy currents to be induced within the conductor. Depending on the frequency of the excitation alternating current as well as the conductivity and relative permeability of the conductor, the eddy current effect may be stronger or weaker. By lowering the frequency of the excitation, alternating current eddy currents will tend to flow at higher depths from the surface of the conductor. If higher frequencies are used (e.g., in the range of several hundreds of kilohertz and above), the depth that eddy currents will flow will be restricted significantly. Based on Lenz's law, if there is no defect present, the induced eddy currents flowing inside the conductor will generate a secondary magnetic field, which will tend to oppose the primary magnetic field created by the excitation coil. In the presence of a defect, the flow of the induced eddy currents will be disturbed and hence the secondary magnetic field will fluctuate, giving rise to changes in the impedance of the sensing coil. These impedance changes can then be related to the size and nature of the defect detected [47].

For several years, the application of eddy current technology was limited for inspection of individual rail welds. More recently, eddy current systems have been developed to perform inspections on rails at speeds of a few meters per minute in order to detect cracks due to RCF. Significant developments in inspection of rails using eddy current technology have been reported in Refs. [38,40,46]. The sensor is pushed by the operator along the rail head who looks for changes in the signal caused by the presence of RCF cracks or wheel burns. It is very important to guide the eddy current probes so that the signals are not influenced and the sensitivity does not fluctuate due to liftoff from the test surface. The rail inspection test situation is especially complex, because the probe has to be positioned at an angle relative to the guiding surface.

3.4.6 Alternating Current Field Measurement Rail Inspection

Alternating current field measurement (ACFM) is an electromagnetic inspection method which is now widely accepted as an alternative to magnetic particle inspection method [49]. Although developed initially for routine inspection of structural welds, the technology has been improved further to cover broader applications across a range of industries. The technique is based on the principle that an alternating current can be induced to flow in a thin skin near the surface of any conductor. By introducing a remote uniform current into an area of the component under test, when there are no defects present, the electrical current will be undisturbed. If a crack is present, the uniform current is disturbed and the current flows around the ends and down the faces of the crack. Because the current is an alternating current, it flows in a thin skin close to the surface and is unaffected by the overall geometry of the component. In contrast to eddy current sensors that are required to be placed at a close (<2 mm) and constant distance from the inspected surface, a maximum operating liftoff of 5 mm is possible without significant loss of signal when using ACFM probes. The incorporated ACFM array has been shaped to conform to the shape of the head of the rail. This allows the application of the ACFM system in both new and worn rails. The inspection across the rail head is carried out by sequentially scanning across the group of sensors enabling the uninterrupted inspection of the rail.

3.4.7 Rail Inspection Using Electromagnetic Transducers

Electromagnetic acoustic transducers (EMATs) may be used to generate and detect ultrasound in an electrically conducting or a magnetic material [50]. This is achieved by passing a large current pulse through an inductive coil in close proximity to a conducting surface in the presence of a strong static magnetic field, often provided by a permanent magnet. The orientation of the magnetic field, geometry of the coil, and physical and electrical properties of the material under investigation have a strong influence on the ultrasound generated within the sample. EMATs have the advantage that they operate without the need for physical coupling or acoustic matching as it is an electromagnetic coupling mechanism that generates the ultrasound within the sample skin depth. This also means that the perturbation that physical coupling causes is insignificant and operation at elevated temperatures is possible. EMATs are therefore suitable for rail inspection.

3.5 Comparison of NDT Techniques

Comparison of the advantages and disadvantages of the NDT methods available to rail and fastening parts discussed earlier is shown in Tables 3.1 and 3.2.

Table 3.1 Advantages and Disadvantages of NDT Methods for Rail and Fastening Parts Inspection

Inspection Technique	Advantages	Disadvantages	Detection Capability
Manual and automated visual	Simple, can be automated, fast, inexpensive, defective ballast, reliable in detecting corrugation	Provides information only regarding the surface of the component, missing parts, nonquantitative	Surface breaking defects, rail head profile missing parts
Liquid penetrant	Simple, high resolution, accurate, very sensitive to small surface-breaking defects, appropriate for weld inspection in rails, applicable to any type of material that is nonporous	Requires surface preparation, access to the component's surface, qualitative, thorough cleaning, no permanent record, only surface-breaking defects detectable	Small surface-breaking defects such as fatigue cracks and corrosion pits
Conventional ultrasonic	Relatively inexpensive unless phased arrays are used, capable of detecting hidden defects and quantifying both hidden and surface-breaking defects, can be applied to any type of material	Local inspection, at high speed can miss surface defects <4 mm as well as internal defects particularly at the rail foot, affected by weather conditions	Internal and surface defects including fatigue cracks and corrosion, no detectable rail foot defects
Laser ultrasonic	Reliable in detecting internal defects, relatively fast	Can be affected by liftoff variations of the sensors, difficult to deploy at high speeds, expensive	Internal and surface defects, rail head, web and foot defects

(Continued)

Table 3.1 (Continued) Advantages and Disadvantages of NDT Methods for Rail and Fastening Parts Inspection

Inspection Technique	Advantages	Disadvantages	Detection Capability
Phased array ultrasonic	Capable of detecting hidden defects and quantifying, can be applied to any type of material	Local inspection, at high speed can miss smaller surface defects, can be affected by liftoff variations of the sensors, quite expensive	Internal and surface defects including fatigue cracks and corrosion, rail head, web and foot defects
Long-range ultrasonics	Relatively fast, capable of detecting large hidden and surface-breaking defects, can be applied to any type of material, can inspect long sections up to several tens of meters in one go	Only simple geometries can be inspected, considerable dead zone, defects need to be relatively large to be detectable, signal-to-noise ratio can be affected by the inspection conditions	Relatively severe corrosion and transverse cracks, surface defects, rail head, web and foot internal defects
Magnetic flux leakage	Fast, sensitive to transverse cracks and corrosion, applicable for surface and hidden defects	Only ferrous materials, defect geometry influences quantification, parallel cracks can be missed, requires good magnetization to avoid underestimation or missed defects, bulky equipment	Surface and hidden corrosion and fatigue cracks, inclusions

(Continued)

Table 3.1 (*Continued*)   Advantages and Disadvantages of NDT Methods for Rail and Fastening Parts Inspection

Inspection Technique	Advantages	Disadvantages	Detection Capability
Eddy currents	Inexpensive, sensitive to microstructural, electric and magnetic properties, sensitive to small defects, applicable to any conductive material, can operate at significant liftoffs	Very liftoff sensitive, inspection penetration depth and resolution dependent on frequency, local inspection, more efficient for surface and near-surface inspection, low resolution in high liftoffs	Surface and near-surface defects (cracks and pitting corrosion), general corrosion, microstructural changes
Alternating current field measurement	Mainly manual system, inexpensive, sensitive to small defects, capable of quantifying depth and length of surface-breaking defects, can be automated, can operate at significant liftoffs	Only surface-breaking defects, local inspection, quantification only possible for fatigue cracks	Surface-breaking defects including pitting corrosion and fatigue cracks
Electromagnetic acoustic transducers	Inexpensive, noncontact, no material limitation as long as it is conductive, can detect both hidden and surface-breaking defect, can be local or long range, can be applied at high temperature, easy to produce specific waves and modes	Low signal-to-noise ratio, sensor requires cooling at high temperatures, bulky sensors, liftoff cannot exceed 2 mm, low-speed hi-rail vehicle (<10 km/h)	Surface and hidden defects including corrosion and fatigue cracks

Table 3.2 Comparison of NDT Methods for Rail and Fastening Parts

Inspection Characteristics	Manual and Automated Visual	Liquid Penetrant	Conventional Ultrasonic	Laser Ultrasonic	Phased Array Ultrasonic	Long Range Ultrasonic	Magnetic Flux Leakage	Eddy Current	Alternating Current Field Measurement	Electromagnetic Acoustic Transducers
						NDT Method				
Detection capability	Limited (surface only)	High (surface only)	High	High	High	Average (large defects only)	High (ferrous only)	High (near surface only)	High (surface only)	High
Detection resolution	Average	High	High	High	High	Low	Average	High	High	Average
Depth estimation	No	No	Yes	Yes	Yes	Yes	Yes	Yes	Yes	Yes
Portability/access	High	High	High	High	High	Low	Average	High	High	Average
Couplant required/surface treatment/surface access	No	Yes	Yes	Yes	Yes	Yes	No	Average (some surface preparation may be required)	No	No
Simplicity	High	High	High	Low	Low	Average	Average	Average	High	Average
Inspection speed	Average	Average	Average	Average	Average	Low	High	High	High	High
Appropriate for use in robotic crawlers (internal or external)	Yes (AVI)	No	Yes	Yes	Yes	No	Yes	Yes	Yes	Yes
Level of training required	Low	Low	High	High	High	High	Average	High	Low	High
Cost	Low	Low	Average	High	High	High	Average	Low	Low	Average

References

1. Signals passed at danger, Office of Rail Regulation, 2011.
2. Melago, C., and C. A. Catastrophic. Train wreck caused when metrolink engineer failed to stop, say rail officials. *Daily News (New York City)*, September 13, 2008.
3. National Transportation Safety Board. Collision of union Pacific railroad train MHOTU-25 with BNSF railway company train MEAP-TUL-126-D with subsequent derailment and hazardous materials release, Macadona, TX, June 28, 2004. NTSB #RAR-06/03, National Transportation Safety Board, Washington, DC, July 2006.
4. Office of Rail Regulation. Train derailment at Hatfield: A final report by the independent investigation board, Office of Rail Regulation, UK, 2006.
5. National Transportation Safety Board. Derailment of Canadian Pacific railway freight train 292-16 and subsequent release of anhydrous ammonia near Minot, North Dakota, January 18, 2002. Railroad Accident Report, RAR-04-01, National Transportation Safety Board, Washington, DC, March 9, 2004.
6. Railway Association of Canada; Association of American Railroads. Algoma Central Railway, British Columbia Railway, Burlington Northern Railroad, CN Rail, CP Rail, Norfolk Southern, and Seaboard System Railroad: Advanced train control systems operating requirements. Railway Association of Canada, Ottawa, Canada; Association of American Railroads, Washington, DC, April 1984.
7. Detmold, P. J. New concepts in the control of train movement. In *Transportation Research Record 1029*, TRB, National Research Council, Washington, DC, 1985, pp. 43–47.
8. FRA, U.S. Department of Transportation. Implementation of positive train control systems. FRA, U.S. Department of Transportation, Washington, DC, 1999.
9. Moore Ede, W. J., A. Polivka, J. Brosseau, Y. Tse, and A. Reinschmidt. Improving enforcement algorithms for communications-based train control using adaptive methods. *Proceedings of the 9th International Heavy Haul Conference—Heavy Haul Innovation and Development*, Shanghai, China, 2009.
10. FRA, U.S. Department of Transportation. Positive train control systems. 49 CFR Parts 229, 234, 235, and 236. FRA, U.S. Department of Transportation, Washington, DC, January 15, 2010.
11. Lee, J.-D., J.-H. Lee, C.-H. Cho, P.-G. Jeong, K.-H. Kim, and Y.-J. Kim. Analysis of moving and fixed autoblock systems for Korean high speed railway. In *Computer in Railways VII*, WIT Press, Southampton, MA, 2000, pp. 842–851.
12. Wendler, E. Influence of ETCS on the capacity of lines. In *Compendium on ERTMS*, Eurail Press, Hamburg, Germany, 2009, pp. 211–223.
13. Smith, M. E., R. R. Resor, and P. Patel. Train dispatching effectiveness with respect to communication-based train control: quantification of the relationship. In *Transportation Research Record 1584*, TRB, National Research Council, Washington, DC, 1997, pp. 22–30.
14. Smith, M. E., and R. R. Resor. The use of train simulation as a tool to evaluate the benefits of the advanced railroad electronics system. *Journal of the Transportation Research Forum*, Vol. 29, No. 1, 1989, pp. 163–168.
15. Smith, M. E., P. K. Patel, R. R. Resor, and S. Kondapalli. Quantification of expected benefits: Meet/pass planning and energy management subsystems of the advanced railroad electronics system (ARES). *Journal of the Transportation Research Forum*, Vol. 30, No. 2, 1990, pp. 301–309.

16. Resor, R. R., M. E. Smith, and P. K. Patel. Positive train control (PTC): Calculating benefits and costs of a new railroad control technology. *Journal of the Transportation Research Forum*, Vol. 44, No. 2, 2005, pp. 77–98.
17. Martland, C. D., and M. E. Smith. Estimating the impact of advanced dispatching systems on terminal performance. *Journal of the Transportation Research Forum*, Vol. 30, No. 2, 1990, pp. 286–300.
18. FRA, U.S. Department of Transportation. Positive train control systems economic analysis. Docket FRA-2006-0132, Notice No. 1 RIN 2130-AC03. FRA, U.S. Department of Transportation, Washington, DC, 2009.
19. Ditmeyer, S. R. Network-centric railroading: Utilizing intelligent railroad systems. *Proceedings of the AREMA C&S Technical Conference*, Louisville, KY, 2006.
20. Daniels, L. E. Track maintenance costs on rail transit properties. TCRP Document 43. TRB, National Research Council, Washington, DC, 2008.
21. Esveld, C. *Modern Railway Track, Delft: MRT-Productions*, 2nd edition, Germany, 2001.
22. International Heavy Haul Association. Guidelines to best practices for heavy haul railway operations: Wheel and rail interface issues. International Heavy Haul Association, Virginia, 2001.
23. Orringer, O., J. Orkisz, and S. Zdzislaw. Residual stress in rails: Effects on rail integrity and railroad economics. Volume II: Theoretical and numerical analyses, Springer, the Netherlands, 1992.
24. Cannon, D. F., K.-O. Edel, S. L. Grassie, and K. Sawle. Rail defects: An overview. *Fatigue and Fracture of Engineering Materials and Structures*, Vol. 26, 2003, pp. 865–887.
25. Grassie, S. L., and J. Kalousek. Rolling contact fatigue of rails: Characteristics, causes and treatments. *Proceedings of the 6th International Heavy Haul Conference*, Cape Town, South Africa, 1997.
26. Murav'ev, V. V., and E. V. Boyarkin. Nondestructive testing of the structural-mechanical state of currently produced rails on the basis of the ultrasonic wave velocity. *Russian Journal of Nondestructive Testing*, Vol. 39, No. 3, 2003, pp. 24–33.
27. Vadillo, E. G., J. A. Tarrago, G. G. Zubiaurre, and C. A. Duque. Effect of sleeper distance on rail corrugation. *Wear*, Vol. 217, 1998, pp. 140–146.
28. Eadie, D.T., J. Kalousek, and K. C. Chiddick. The role of high positive friction (HPF) modifier in the control of short pitch corrugation and related phenomena. *Contact Mechanics Conference*, Tokyo, Japan, 2000.
29. Jeong, D.Y., Progress in rail integrity research. Report DOT/FRA/ORD-01/18, Federal Railroad Administration, Department of Transportation, Washington, DC, 2001.
30. Cannon, D. F., and H. Pradier. Rail rolling contact fatigue research by the European Rail Research Institute. *Wear*, Vol. 191, 1996, pp. 1–13.
31. Grassie, S., P. Nilsson, K. Bjurstrom, A. Frick, and L. G. Hansson. Alleviation of rolling contact fatigue on Sweden's heavy haul railway. *Wear*, Vol. 253, 2002, pp. 42–53.
32. Pandrol. The future of rail fastenings, www.pandrol.com.
33. Campos-Castellanos, C., D1.1 Project requirements and specifications preparation of materials, Long range inspection and condition monitoring of rails using guided waves, MonitoRail, Grant agreement no: 262194, Cambridge, UK, 2011.
34. Sustainable bridges, http://www.sustainablebridges.net/, 2014.
35. Economic Commission for Europe, Inland Transport Committee, Railway Tunnels in Europe and North America, Informal document no. 7, United Nations Economic and Social Council (UNECE), May 2002.

36. Papaelias, M.P., R. C., and C. L. Davis. A review on non-destructive evaluation of rails: State-of-the-art and future development. *Journal of Rail and Rapid Transit,* Vol. 222, No. 4, 2008, pp. 367–384.

37. Campos-Castellanos, C., Y. Gharaibeh, P. Mudge, and V. Kappatos. The application of long range ultrasonic testing (LRUT) for examination areas with difficult access on railway tracks. *5th IET Conference on Railway Condition Monitoring and Non-Destructive Testing,* Derby, November 29–30, 2011.

38. Krull, R., H. M. Thomas, R. Pohl, and S. Rühe. Eddy current detection of head-checks on the gauge corner of rails: Recent results. *Conference on Railway Engineering,* London, 2003.

39. Pohl, R., R. Krull, and R. Meierhoffer. A new eddy current instrument in a grinding train. *ECNDT Conference—Poster 178,* Berlin, Germany, September 2006.

40. Moustakidis, S., V. Kappatos, P. Karlsson, C. Selcuk, T.-H. Gan, and K. Hrissagis. An intelligent methodology for railways monitoring using ultrasonic guided waves. *Journal of Nondestructive Evaluation,* Vol. 33, No. 4, 2014, pp. 694–710.

41. American Society of Non-Destructive Testing. Introduction to non-destructive testing, American Society of Non-Destructive Testing Presentation. www.nde-ed.org/GeneralResources/IntroToNDT/Intro_to_NDT.ppt., 2011.

42. Transportation Technology Center. *12th Annual AAR Research Review,* Pueblo, CO, 2007.

43. Rose, J. L. Standing on the shoulders of giants: An example of guided wave inspection. *Materials Evaluation,* Vol. 60, No. 1, 2002, pp. 53–59.

44. Chuck, H., *Handbook of Nondestructive Evaluation,* 2nd edition, McGraw-Hill, New York, 2012.

45. Rose, J. L., M. J. Avioli, and W. J. Song. Application and potential of guided wave rail inspection. *Insight: Non-Destructive Testing and Condition Monitoring,* Vol. 44, No. 6, 2002, pp. 353–358.

46. Wilson, J. W., and G. Y. Tian. 3D magnetic field sensing for magnetic flux leakage defect characterisation. *Insight,* Vol. 48, No. 6, 2006, p. 357.

47. Mcmaster, R. C., and M. L. Mester Electromagnetic Testing: Eddy Current, Flux Leakage, and Microwave Nondestructive Testing (Nondestructive Testing Handbook) Hardcover – September, 1986.

48. Thomas, H. M., M. Junger, H. Hintze, R. Krull, and S. Rühe. Pioneering inspection of railroad rails with eddy currents. *Proceedings of the 15th WCNDT,* Rome, Italy 2000.

49. Paul, E. M. *Introduction to Non-Destructive Testing: A Training Guide,* John Wiley & Sons, 2005, pp. 124–126.

50. Thompson, R. B. Physical principles of measurements with EMAT transducers. In *Ultrasonic Measurement Methods: Physical Acoustics. Vol. XIX,* Edited by Thurston, R. N., and A. D. Pierce, Academic Press, New York, 1990.

Chapter 4

Modeling of the Wireless Channels in Underground Tunnels for Communications-Based Train Control Systems

Hongwei Wang, F. Richard Yu, Li Zhu, and Tao Tang

Contents

4.1 Introduction 66
4.2 Real Field CBTC Channel Measurements 67
    4.2.1 Measurement Equipment 67
    4.2.2 Measurement Scenario 68
4.3 Finite-State Markov Chain Channel Model 69
    4.3.1 FSMC Model 69
    4.3.2 Determine the SNR-Level Thresholds of the FSMC Model 72
    4.3.3 Determine the Distribution of SNR 74
4.4 Real Field Measurement: Results and Discussions 76
4.5 Conclusion 79
References 79

4.1 Introduction

Building a train–ground wireless communication system for communications-based train control (CBTC) is a challenging task. As urban rail transit systems are mostly deployed in underground tunnels, there are a large amount of reflections, scattering, and barriers that severely affect the propagation performance of wireless communications. Moreover, due to the available commercial off-the-shelf equipment, wireless local area networks (WLANs) are often adopted as the main method of train–ground communications for CBTC systems. However, most of the current IEEE 802.11 WLAN standards are not originally designed for the high-speed environment in tunnels [1]. Furthermore, the fast movement of trains will cause frequent handoffs between WLAN access points (APs), which can severely affect the CBTC performance.

Modeling the channels of urban rail transit systems is very important to design the wireless networks and evaluate the performance of CBTC systems. There are some previous works on radio wave propagation in urban rail transit systems. A path loss model of tunnel channels is given in [2], which describes the characteristics of the large-scale fading. The authors of [3] present the propagation characteristics based on real environment measurements in Madrid subway. A two-layer multistate Markov model is presented in [4] for modeling a 1.8 GHz channel in urban Taipei city. Based on the Winner II physical layer channel model parameters, the authors of [5] propose a channel model for high-speed railway.

Although some excellent works have been done on modeling channels, most of them do not consider the unique characteristics of CBTC systems, such as high mobility speed, deterministic moving direction, and accurate train location information. In this chapter, we develop a finite-state Markov channel (FSMC) model for tunnel channels in CBTC systems. FSMC models have been widely accepted in the literature as an effective approach to characterize the correlation structure of the fading process, including 1.8 GHz narrow-band channels [4], high-speed railway channels [5], satellite channels [6], indoor channels [7], Rayleigh fading channels [8], Rician fading channels [9], and Nakagami fading channels [10]. Using FSMC models, a variety of analytical results of system performance can be derived, including channel capacity [11], throughput [12], and packet error distribution [13].

To the best of our knowledge, FSMC models for tunnel channels in CBTC systems have not been studied in previous works. Therefore, there is a strong motivation to develop an FSMC model for tunnel channels in CBTC systems. Some distinct features of the proposed channel model are as follows:

- The proposed FSMC model is based on real field CBTC channel measurements obtained from the business operating Beijing Subway Changping Line.
- Unlike most existing channel models, which do not use train location information, the proposed FSMC channel model takes train locations into account to have a more accurate channel model.

- The distance between the transmitter and the receiver is divided into intervals, and the FSMC model is applied in each interval.
- Lloyd–Max technique [14] is used to determine the signal-to-noise ratio (SNR)-level boundaries in the proposed FSMC model.
- The accuracy of the proposed FSMC model is illustrated by the simulation results generated from the model and the real field measurement results. The effects of different parameters are also discussed.

The rest of this chapter is organized as follows: Section 4.2 describes the real field measurement configuration and scenario. Section 4.3 introduces the FSMC model. Then, Section 4.4 presents the real field measurement results and discussions. Finally, Section 4.5 concludes the chapter.

4.2 Real Field CBTC Channel Measurements

The objective of the real field CBTC channel measurements is to get the real field data of WLAN propagation in tunnels under real conditions of the subway line, which will be used to build an FSMC model. In this section, we present the measurement equipment and the measurement scenario in our real field CBTC channel measurements.

With this objective, the preparation of the measurements consists of two parts as follows:

1. We need to make sure that the configuration of the measurements is the same as business operating subway lines, including the choice of antennas, the location of antennas, and the settings of the transmitter and the receiver.
2. We need to develop a measurement method to map channel data, including the signal strength and SNR, to the location of the receiver, which will be used in our research that takes train locations into account to have a more accurate channel model.

4.2.1 Measurement Equipment

Two Cisco 3200 routers are used in our measurements. One is set as the AP, whereas the other one is set as the mobile station (MS). Both of them are set to work at the frequency of 2.412 GHz, which is also called channel 1. The output power of the AP is set as 30 dBm. The AP is located on the wall of the tunnel, whereas the MS is located on the measurement vehicle. The transmitting antenna is a Yagi antenna connected with the AP, which is directional and vertically polarized. The half power beam width (HPBW) is 30° and the gain of Yagi antenna is 13.5 dBi. In addition, the Shark-fin antenna is applied as the receiving antenna connected with the MS, which is also directional and vertically polarized. The HPBW is 40° and the gain of Shark-fin antenna is 10 dBi.

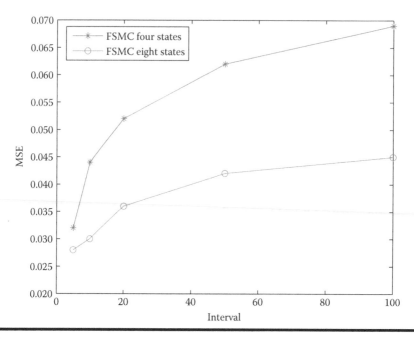

Figure 4.1 Measurement equipment used in the real field CBTC channel measurements.

The location of the receiver is obtained through a velocity sensor installed on the wheel of the measurement vehicle, which can detect the real-time velocity, and the resolution of position is millimeter per second. Figure 4.1 shows the measurement equipment used in our real field measurements.

4.2.2 Measurement Scenario

The measurements were performed in Beijing Subway Changping Line, where the cross section of the tunnel is rectangular. The cross section of the tunnel and the locations of antenna are shown in Figure 4.2. The height of the tunnel is 4.91 m and the width is 4.4 m. The transmitting antenna is located 0.15 m below the tunnel roof. The receiving antenna is set on the top of a measurement vehicle. As the threshold of the receiver is −90 dBm, the coverage of one AP is about 0–500 m, which is also the experimental zone in our measurements. The tunnel where we performed the measurement is a section of straight tunnel. Figure 4.3 shows the cross section of the tunnel near Nanshao station of Beijing Subway Changping Line, the Shark-fin antenna, the Yagi antennas, and the AP set on the wall. We performed measurements in the tunnel of Beijing Subway Changping Line for 20 times so that enough data can be captured.

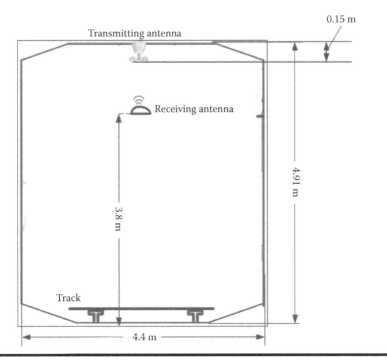

Figure 4.2 Tunnel section and deployment of antennas in the measurement.

4.3 Finite-State Markov Chain Channel Model

To capture the characteristics of tunnel channels in CBTC systems, we define channel states according to the different SNR levels received and use an FSMC to track the state variation. In this section, we first describe the FSMC model, followed by the determination of key model parameters, including SNR levels and SNR distribution. Table 4.1 illustrates the notions of symbols used in this chapter.

4.3.1 FSMC Model

The SNR range of the received signal can be partitioned into N nonoverlapping levels with thresholds $\{\Gamma_n, n = 0,1,2,3,...,N\}$, where Γ_0 and Γ_N can be measured. The time axis is divided into slots of equal duration. Let γ_k denote the channel state in time slot k. $\gamma_k = s_n$ when the SNR of the received signal belongs to the range (Γ_{n-1}, Γ_n). Then the received SNR can be modeled as a random variable γ evolving according to a finite-state Markov chain, and the transition probability $p_{n,j}$ can be shown as follows:

$$p_{n,j} = P_r\{\gamma_{k+1} = s_n \mid \gamma_k = s_j\} \tag{4.1}$$

Figure 4.3 **(a) Tunnel where we performed the measurements in Beijing Subway Changping Line. (b) Shark-fin antenna located on the measurement vehicle. (c) Yagi antenna. (d) AP set on the wall.**

where:
$$k = 1, 2, 3, \ldots$$
$$n, j \in \{1, 2, \ldots, N\}$$

Based on the measurement data, we observe that most transitions within a Markov chain are between adjacent states. Therefore, we assume that each state can only transit to the adjacent states, which means $p_{n,j} = 0$, if $|n - j| > 1$. With the definition, we can define a state transition probability matrix $\mathbf{P}$ with elements $p_{n,j}$.

Table 4.1 Notions of Symbols

γ_k	The channel state in time slot k
N	The number of SNR levels
L	The number of distance intervals
Γ_n	The threshold of the nth level of SNR
s_n	The channel state n
$p_{n,j}$	The transition probability from state s_j to state s_n
$\mathbf{P}^l$	The transition probability matrix in the l th interval
s_n^l	The channel state n in the l th interval
$p_{n,j}^l$	The transition probability from state s_j^l to state n
a_n^l	The number of times state s_n^l appears
$a_{n,j}^l$	The number of times that states s_j^l transits to state s_n^l
$\tilde{\Gamma}_n$	The quantized value of SNR in the range $(\Gamma_{n-1}\,\Gamma_n)$
L_m	The maximized value of the likelihood function
η	The number of parameters of the statistical model
n_s	The number of channel samples

Due to the effect of large-scale fading, the amplitude of SNR depends on the distance between the transmitter and the receiver. It is obvious that the SNR is usually high when the receiver is close to the transmitter, whereas it is low when the receiver is far away from the transmitter. As a result, the transition probability from the high channel state to the low channel state is different when the receiver is near or far away from the transmitter, which means that the Markov state transition probability is related to the location of the receiver. Therefore, only one state transition probability matrix, which is independent of the location of the receiver, may not accurately model the tunnel channels. Thus, we divide the tunnel into L intervals and one state transition probability matrix is generated for each interval. Specifically, $\mathbf{P}^l, l \in \{1,2,...,L\}$ is the state transition probability matrix corresponding to the l th interval, and the relationship between the transition probability and the location of the receiver can be built. Then, $p_{n,j}^l$ is the state transition probability from state s_n to state s_j in the l th interval. And the state n and the state j in the l th interval are denoted as s_n^l and s_j^l, respectively.

Consequently, the state probabilities and the state transition probabilities can be defined as follows:

$$p_n^l = P_r^l \{\gamma_k^l = s_n^l\}$$

$$p_{n,j}^l = P_r^l \{\gamma_{k+1}^l = s_n^l \mid \gamma_k^l = s_j^l\}$$

$$p_{n,j}^l = 0, \text{ if } |n-j| > 1 \tag{4.2}$$

$$\sum_{i=1}^N p_{n,j}^l = 1, \forall n \in \{1,2,3,...,N\}$$

where:

p_n^l is the probability of being in state n in the l th interval

γ_k^l is the SNR level in time slot k in the l th interval

Based on the measurement results, we can determine the value of the state probability p_n^l and the state transition probability $p_{n,j}^l$:

$$p_n^l = \frac{a_n^l \{\gamma_k^l = s_n^l\}}{\sum_{n=1}^N a_n^l \{\gamma_k^l = s_n^l\}} \tag{4.3}$$

where $a_n^l \{\gamma_k^l = s_n^l\}$ is the number of times state s_n appears in the lth interval.

$$p_{n,j}^l = \frac{a_{n,j}^l \{\gamma_{k+1}^l = s_n^l \mid \gamma_k^l = s_j^l\}}{\sum_{j=1}^N a_{n,j}^l \{\gamma_{k+1}^l = s_n^l \mid \gamma_k^l = s_j^l\}} \tag{4.4}$$

where $a_{n,j}^l \{\gamma_{k+1}^l = s_n^l \mid \gamma_k^l = s_j^l\}$ is the number of times that state j transits to state n in the l th interval.

4.3.2 Determine the SNR-Level Thresholds of the FSMC Model

Determining the thresholds of SNR levels is the key factor that affects the accuracy of the FSMC model. There are many methods to select the SNR-level boundaries, among which the equiprobable partition method is frequently used in previous works [9,10]. As nonuniform amplitude partitioning can be useful to obtain more accurate estimates of system performance measures [15], we choose the Lloyd–Max technique [14] instead of the equiprobable method to partition the amplitude of SNR in this chapter. Lloyd–Max is an optimized quantizer, which can decrease the distortion of scalar quantization.

First, a distortion function D is defined as follows:

$$D = \sum_{n=1}^{N} \int_{\Gamma_{n-1}}^{\Gamma_n} f(\tilde{\Gamma}_n - \gamma) p(\gamma) d\gamma \qquad (4.5)$$

where:

$\tilde{\Gamma}_n$ is the quantized value of SNR whose amplitude is in the range $(\Gamma_{n-1} \ \Gamma_n)$
$f(x)$ is the error criterion function
$p(\gamma)$ is the probability distribution function of SNR

The distortion function can be minimized through optimally selecting $\tilde{\Gamma}_n$ and Γ_n.

Then, the necessary conditions for minimum distortion are obtained by differentiating D with respect to Γ_n and $\tilde{\Gamma}_n$. The result of this minimization is a pair of equations [16]:

$$f(\tilde{\Gamma}_n - \Gamma_n) = f(\tilde{\Gamma}_{n+1} - \Gamma_n) \qquad (4.6)$$

$$\int_{\Gamma_{n-1}}^{\Gamma_n} f'(\tilde{\Gamma}_n - \gamma) p(\gamma) d\gamma = 0 \qquad (4.7)$$

The error criterion function $f(x)$ is often taken as x^2 [16]. As a result, Equations 4.6 and 4.7 become

$$\Gamma_n = \frac{\tilde{\Gamma}_n + \tilde{\Gamma}_{n+1}}{2} \qquad (4.8)$$

$$\int_{\Gamma_{n-1}}^{\Gamma_n} (\tilde{\Gamma}_n - \gamma) p(\gamma) d\gamma = 0 \qquad (4.9)$$

As mentioned above, we partition the amplitude of SNR into N levels, and there are $N+1$ corresponding thresholds $\{\Gamma_n, n = 0,1,2,3,...,N\}$. Generally, the first and last thresholds are known, which are denoted by the minimum and maximum measurement values of SNR, respectively. Furthermore, the Lloyd–Max algorithm is used to divide 2^r levels, which means $N = 2^r, r = 1,2,3,...,$ and N is an even number. As a result, as Γ_0 and Γ_N are known, $\Gamma_{N/2}$ can be obtained from Equation 4.9. Then, $\Gamma_{N/4}$ and $\Gamma_{3N/4}$ can also be calculated according to the new variable $\Gamma_{N/2}$, when r is larger than 2. With the process being repeated, all elements of $\{\Gamma_n\}$ can be obtained as follows:

$$\int_{\Gamma_a}^{\Gamma_b} \left(\tilde{\Gamma}_{\frac{b+a}{2}} - \gamma \right) p(\gamma) d\gamma = 0 \tag{4.10}$$

$$b > a \text{ and } b \in \{2,3,...,N\}, a \in \{1,2,3,...,N-1\}$$

where:

$p(\gamma)$ is the probability distribution function of SNR

According to the calculated $\{\Gamma_n\}$, combined with Equations 4.8 and 4.9, we can update the value of $\{\Gamma_n\}$ until the value of D is the minimum, and the optimal thresholds of the SNR levels can be obtained. As $p(\gamma)$ is still unknown, we should discuss the distribution of SNR in Section 4.3.3 according to the real field measurement data, which is the last step to obtain the thresholds of SNR levels.

4.3.3 Determine the Distribution of SNR

Deriving the distribution of SNR is the crucial step of partitioning the levels of SNR. In fact, there are some classic models to describe the distribution of signal strength, such as Rice, Rayleigh, and Nakagami, and then the corresponding models of SNR can also be obtained [17]. We first derive the distribution of the signal strength in order to determine the distribution model of SNR. According to [17] and [18], voltage/meter (v/m) is usually used when studying the distribution of small-scale fading. Therefore, we convert dBm to the linear unit v/m following [19].

The Akaike information criterion (AIC) is often used to get the approximate distribution model of the signal strength [20]. It is a measure of the relative goodness of fit of a statistical model. It was developed by Hirotsugu Akaike, under the name of an information criterion, and was first published in 1974. The general case of AIC is

$$\text{AIC} = -2\ln L_m + 2\eta \tag{4.11}$$

where:

η is the number of parameters of the statistical model
L_m is the maximized value of the likelihood function for the estimated model

In fact, according to the relationship of η and the number of samples n_s, AIC needs to be changed to AIC with a correction (AICc) when $n_s / \eta < 40$ [20].

$$\text{AICc} = \text{AIC} + \frac{2\eta(\eta+1)}{n_s - \eta - 1} \tag{4.12}$$

In this chapter, AICc is adopted to estimate the model of the signal strength distribution instead of the classic AIC. In practice, one can compute AICc for each of the candidate models and select the model with the smallest value of AICc. The candidate models in this chapter include Rice, Rayleigh, and Nakagami.

Because our channel model is related to the distance between the transmitter and the receiver, the tunnel should be divided into intervals. Assume that there are L intervals, and we apply AICc for each candidate model in every interval. As a result, we can select the most appropriate model based on the frequency of the minimum AICc value of different candidate models. In order to obtain enough data for each interval and ensure the accuracy of the model, we set the length of each interval as 40 wavelengths of WLANs [18], and then there are 100 intervals. Based on the frequencies of AICc of different distributions in the real field measurements, we observe that the Nakagami distribution provides the best fit compared to Rayleigh and Rician distributions as shown in Figure 4.4. As a result, we can define $p(\gamma)$ as the Nakagami distribution.

After the distribution of the signal strength is obtained, according to [17], we can derive the distribution of SNR:

$$p(\gamma_l) = \frac{\mu_l^{\mu_l} \gamma_l^{\mu_l - 1}}{\bar{\gamma}_l^{\mu_l} \Gamma(\mu_l)} e^{(-\mu_l \gamma_l / \bar{\gamma}_l)} \tag{4.13}$$

where:

γ_l is the SNR of the received signal in the l th interval

$\bar{\gamma}_l$ is the mean of SNR in the lth interval

μ_l is the fading factor of Nakagami distribution in the l th interval

$\Gamma(\cdot)$ is the gamma function

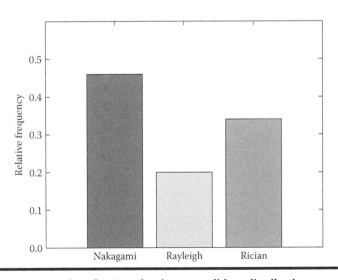

Figure 4.4   Frequencies of AICc selecting a candidate distribution.

In fact, μ_l can be calculated when applying AICc through the maximum likelihood estimator for each interval.

4.4 Real Field Measurement: Results and Discussions

In this section, we compare our FSMC model with real field test results to illustrate the accuracy of the model. The effects of different parameters in the proposed model are discussed. The number of states in our model is first set as four. We also use eight states to study the effects of the number of states on the accuracy of the proposed model. In order to obtain the effects of distance intervals on the model, we choose the intervals as 5, 10, 20, 50, and 100 m. We perform measurements in the tunnel of Beijing Subway Changping Line for 20 times so that enough data can be captured. The accuracy of the FSMC model is verified through another set of measurement data.

Based on the measurement data, Equations 4.8–4.10 and 4.13, we derive the thresholds $\{\Gamma_n, n = 0,1,2,...,N\}$ of SNR in each distance interval. Tables 4.2 and 4.3 demonstrate the thresholds of the SNR levels at the location of 100 m for different intervals, where we divide SNR into four and eight levels. As the distance intervals are different, the range of SNR is different and it brings different thresholds, which can provide a more accurate model.

After we get the thresholds, we can get the state probabilities and the state transition probabilities from the real field data. Table 4.4 illustrates the state transition probabilities of the FSMC model and the measurement data at the same location (35–40 m), when there are four states and the distance interval is 5 m. We can observe that the sum of the transition probability of each channel state does not equal to 1. This is because, in the measurement data, there are some state transitions that do not happen in adjacent states, such as transitions from state 1 to state 3. However, in our FSMC models, for the sake of simplicity, we assume that states can

Table 4.2 Thresholds of SNR Levels (Four Levels) at the Location of 100 m for Different Intervals

Interval	5 m	10 m	20 m	50 m	100 m
Range	[95,100]	[90,100]	[80,100]	[50,100]	[0,100]
First threshold	24	22	22	22	22
Second threshold	27.98	26.90	27.73	29.53	33.22
Third threshold	32.03	31.44	32.89	35.39	44.01
Fourth threshold	36.31	36.02	38.14	41.22	57.00
Fifth threshold	41	41	44	48	78

Table 4.3 Thresholds of SNR Levels (Eight Levels) at the Location of 100 m for Different Intervals

Interval	5 m	10 m	20 m	50 m	100 m
Range	[95,100]	[90,100]	[80,100]	[50,100]	[0,100]
First threshold	24	22	22	22	22
Second threshold	25.99	24.53	25.00	26.16	27.87
Third threshold	27.98	26.90	27.73	29.50	33.22
Fourth threshold	29.99	29.18	30.33	32.50	38.50
Fifth threshold	32.03	31.43	32.89	35.38	44.01
Sixth threshold	34.13	33.69	35.47	38.25	50.04
Seventh threshold	36.31	36.01	38.14	41.22	57.00
Eighth threshold	38.59	38.43	40.95	44.41	65.68
Ninth threshold	41	41	44	48	78

Table 4.4 State Transition Probabilities of the FSMC Model and the Measurement Data with Four States and 5 m Interval at the Location (35–40 m)

	FSMC Model			Measurement Data		
	$p_{k,k-1}$	$p_{k,k}$	$p_{k,k+1}$	$p_{k,k-1}$	$p_{k,k}$	$p_{k,k+1}$
$k = 1$	–	0.91	0.08	–	0.91	0.08
$k = 2$	0.043	0.86	0.086	0.041	0.86	0.09
$k = 3$	0.024	0.85	0.12	0.024	0.85	0.11
$k = 4$	0.023	0.96	–	0.023	0.97	–

only transit to the adjacent states. Therefore, in Table 4.4, we only consider state transitions between adjacent states. Consequently, the sum of the transition probability of each state of the measurement data does not equal to 1. Nevertheless, it is very close to 1, which means that our assumption is realistic for practical tunnel channels in CBTC systems.

Figure 4.5 shows the simulation results generated from our FSMC model and the experimental results from real field measurements. We can observe the great agreement between them. Next, we derive the mean square error (MSE) to measure the degrees of approximation, shown in Figure 4.6, where the *y*-axis is the

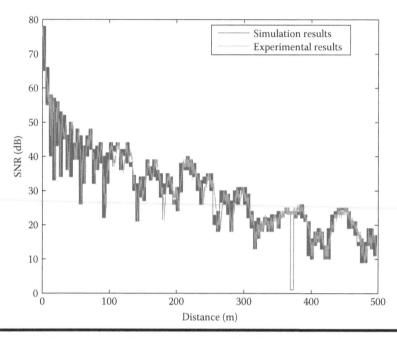

Figure 4.5   Simulation results generated from the FSMC model and experimental results from real field measurements.

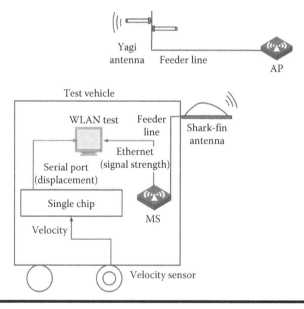

Figure 4.6   MSE between the FSMC model and the experimental data with four states and eight states.

MSE between the FSMC results and the measurement results, and the x-axis is the interval distance (5, 10, 20, 50, and 100 m). As we can see from Figure 4.6, when the distance interval increases, the MSE also increases, which means that the accuracy of the model decreases. Moreover, we can also observe that the MSE of the FSMC model with four states is larger than that with eight states. The number of states in the FSMC model plays a key role in the accuracy. Nevertheless, when the distance interval is 5 m, the MSE difference is small for the four-state FSMC model (0.032) and eight-state FSMC model (0.028). From this figure, we can see that the FSMC model with four states and 5 m distance interval can provide an accurate enough channel model for tunnel channels in CBTC systems.

4.5 Conclusion

Modeling the tunnel wireless channels of urban rail transit systems is important in designing the wireless networks and evaluating the performance of CBTC systems. In this chapter, we have proposed an FSMC model for tunnel channels in CBTC systems. As the train location is known in CBTC systems, the proposed FSMC channel model takes train locations into account to have a more accurate channel model. The distance between the transmitter and the receiver is divided into intervals, and an FSMC model is designed in each interval. The accuracy of the proposed model has been illustrated by the simulation results generated from the proposed model and the real field measurements. In addition, we have shown that the number of states and the distance interval have impacts on the accuracy of the proposed FSMC model.

References

1. L. Zhu, F. R. Yu, B. Ning, and T. Tang. Cross-layer handoff design in MIMO-enabled WLANs for communication-based train control (CBTC) systems. *IEEE J. Sel. Areas Commun.*, 30(4):719–728, 2012.
2. Y. P. Zhang. Novel model for propagation loss prediction in tunnels. *IEEE Trans. Veh. Tech.*, 52(5):1308–1314, 2003.
3. K. Guan, Z. Zhong, J. I. Alonso, and C. Briso-Rodriguez. Measurement of distributed antenna systems at 2.4 GHz in a realistic subway tunnel environment. *IEEE Trans. Veh. Tech.*, 61(2):834–837, 2012.
4. H.-P. Lin and M.-J. Tseng. Two-layer multistate Markov model for modeling a 1.8 GHz narrow-band wireless propagation channel in urban Taipei city. *IEEE Trans. Veh. Tech.*, 54(2):435–446, 2005.
5. S. Lin, Z. Zhong, L. Cai, and Y. Luo. Finite state Markov modelling for high speed railway wireless communication channel. In *Proceedings of the IEEE Globecom*, Anaheim, CA, 2012.
6. F. Babich, G. Lombardi, and E. Valentinuzzi. Variable order Markov modeling for LEO mobile satellite channels. *Electron. Lett.*, 35(8):621–623, 1999.

7. F. Babich and G. Lombardi. A measurement based Markov model for the indoor propagation channel. In *Proceedings of the IEEE VTC*, vol. 1, Phoenix, AZ, May 1997.

8. H. S. Wang and N. Moayeri. Finite-state Markov channel—A useful model for radio communication channels. *IEEE Trans. Veh. Tech.*, 44(1):163–171, 1995.

9. C. Pimentel, T. H. Falk, and L. Lisboa. Finite-state Markov modeling of correlated Rician-fading channels. *IEEE Trans. Veh. Tech.*, 53(5):1491–1501, 2004.

10. C. D. Iskander and P. T. Mathiopoulos. Fast simulation of diversity Nakagami fading channels using finite-state Markov models. *IEEE Trans. Broadcasting*, 49(3):269–277, 2003.

11. A. J. Goldsmith and P. P. Varaiya. Capacity, mutual information, and coding for finite-state Markov channels. *IEEE Trans. Inform. Theory*, 42(3):868–886, 1996.

12. A. Chockalingam, M. Zorzi, L. B. Milstein, and P. Venkataram. Performance of a wireless access protocol on correlated Rayleigh-fading channels with capture. *IEEE Trans. Commun.*, 46(5):644–655, 1998.

13. F. Babich, O. E. Kelly, and G. Lombardi. Generalized Markov modeling for flat fading. *IEEE Trans. Commun.*, 48(4):547–551, 2000.

14. S. Lloyd. Least squares quantization in PCM. *IEEE Trans. Inform. Theory*, 28(2):129–137, 1982.

15. P. Sadeghi, R. Kennedy, P. Rapajic, and R. Shams. Finite-state Markov modeling of fading channels—A survey of principles and applications. *IEEE Signal Proc. Mag.*, 25(5):57–80, 2008.

16. J. G. Proakis. *Digital Communications*. McGraw-Hill, New York, 1995.

17. M. K. Simon and M.-S. Alouini. *Digital Communication over Fading Channels*. John Wiley & Sons, New York, 2005.

18. S. Wyne, A. P. Singh, F. Tufvesson, and A. F. Molisch. A statistical model for indoor office wireless sensor channels. *IEEE Trans. Wireless Commun.*, 8(8):4154–4164, 2009.

19. T. S. Rappaport. *Wireless Communications Principles and Practice*, 2nd Edition. Publishing House of Electronics Industry, Beijing, China, 2008.

20. K. P. Burnham and D. R. Anderson. *Model Selection and Multi-Model Inference: A Practical Information-Theoretic Approach*. Springer, Berlin, Germany, 2002.

Chapter 5

Modeling of the Wireless Channels with Leaky Waveguide for Communications-Based Train Control Systems

Hongwei Wang, F. Richard Yu, Li Zhu, and Tao Tang

Contents

5.1 Introduction 82
5.2 Leaky Waveguide in CBTC Systems 82
    5.2.1 Overview of CBTC Radio Channel with Leaky Waveguide 82
5.3 Measurement Campaign 83
5.4 Modeling of the CBTC Radio Channel with Leaky Waveguide 84
    5.4.1 Measurement Results... 84
    5.4.2 Determination of the Path Loss Exponent 85
    5.4.3 Determination of the Small-Scale Fading 88
5.5 Conclusion 91
References 91

5.1 Introduction

As urban rail transit systems are deployed in a variety of environments (subway tunnels, viaducts, etc.), there are different wireless network configurations and propagation media. For the viaduct scenarios, leaky rectangular waveguide is a popular approach, as it can provide better performance and stronger anti-interference ability than the free space [1]. For example, leaky waveguide has been applied in Beijing Subway Yizhuang Line. In addition, due to the available commercial off-the-shelf equipment, wireless local area networks (WLANs) are often adopted as the main method of train–ground communications for communications-based train control (CBTC) systems [2].

For general applications, leaky waveguide is taken as a leaky wave antenna with the length of several operation wavelengths. However, in CBTC systems, the length of leaky waveguide is several hundred meters and the distance between the leaky waveguide and the receiving antenna is short (about 30 cm). Due to the specificity of the CBTC application, there are only a few research works about leaky waveguide in CBTC systems. The author of [1] gives a description of leaky waveguide used in CBTC systems, and the advantages of leaky waveguide are demonstrated by comparisons with natural propagation. The characteristics of leaky waveguide are shown in [3] through laboratory measurements.

In this chapter, based on the measurement results in Beijing Subway Yizhuang Line, we use the polynomial fitting and equivalent magnetic dipole method to build the path loss model. In addition, the Akaike information criterion with a correction (AICc) is applied to determine the distribution model of the small-scale fading. The proposed path loss model of the channel with leaky waveguide in CBTC systems is linear, and the path loss exponent can be approximated by the transmission loss of leaky waveguide. We show that the small-scale fading follows log-normal distribution, which is often referred to as the distribution model of the small-scale fading in body area communication propagation channels [4,5]. In addition, the corresponding parameters of log-normal distribution μ_{dB} and σ_{dB} are also determined from the measurement results.

The rest of this chapter is organized as follows: Section 5.2 describes an overview of CBTC systems and the application scenario of leaky waveguide in CBTC systems. Section 5.3 discusses the real field measurement configuration and scenario. Then, Section 5.4 presents the path loss model and the small-scale fading model. Finally, Section 5.5 concludes the chapter.

5.2 Leaky Waveguide in CBTC Systems

5.2.1 Overview of CBTC Radio Channel with Leaky Waveguide

Generally speaking, the radio waves of CBTC systems are often transmitted in the free space, especially in tunnels. However, in the viaduct scenarios, the performance of wireless communication could be affected by the interference from

Figure 5.1 Leaky waveguide applied in viaduct scenarios of Beijing Subway Yizhuang Line.

other wireless devices in surrounding buildings. There are periodic transverse slots in the wide wall of leaky waveguide, which can provide stable signals and anti-interference ability. As a result, leaky waveguide has gradually been applied as the propagation medium in CBTC systems. It can be deployed along the track, as shown in Figure 5.1. Considering the unique characteristics of leaky waveguide in CBTC systems, we propose a model of the channel with leaky waveguide to facilitate the design and evaluation of the performance of CBTC systems in this chapter.

5.3 Measurement Campaign

Our measurements were performed at the section from Rongjingdong Station to Tongjinan Station of Beijing Subway Yizhuang Line under real operation conditions. Two sets of Cisco WLAN devices are used, one of which is set as the access point (AP) and the other one is set as the mobile station (MS). Both of them are set to work at the frequency of 2.412 GHz, which is also called channel 1. The output power of the AP taken as the transmitter is set as 30 dBm. Through a coupling unit, the AP is connected with the leaky waveguide. The MS taken as the receiver is located on a measurement vehicle with a panel antenna to receive the signals leaked from the leaky waveguide. The gain of the antenna is 11 dBi and the beam width is 30°. Due to the limits of subway line and the restricted conditions of trains, especially the Bogie, the distance between the antenna and the leaky waveguide is within the scope of 300–500 mm. In our measurements, the receiving antenna was fixed on the measurement vehicle 320 mm above the leaky waveguide in order to capture the channel samples in the near field of the receiving antenna. The length of one section of leaky waveguide is about 300 m.

The location of the receiver is obtained through a velocity sensor installed on the wheel of the measurement vehicle, which can detect the real-time velocity, and

Figure 5.2   Measurement equipment used in the CBTC channel measurements.

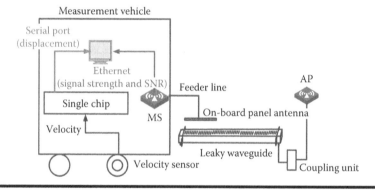

Figure 5.3   Measurement scenario.

the resolution of position is millimeter per second. The measurement vehicle moved along the tracks with the velocity of 70 km/h. Figure 5.2 shows the measurement equipment used in our measurements. Figure 5.3 shows the measurement scenario.

5.4 Modeling of the CBTC Radio Channel with Leaky Waveguide

5.4.1 Measurement Results

We have performed 20 measurements to obtain the statistical characteristics of the channel with leaky waveguide. We use the polynomial fittings of degree 1 and degree 2 to process the experimental data. In the data processing, we find that the quadratic coefficients of polynomial fittings of degree 2 are very small (about 10^{-4}), which means that the quadratic coefficients of the polynomials can be ignored. Hence, the fading channel with the leaky waveguide can be approximated as a linear model according to the polynomial fitting results.

Some classic channel models have been proposed for different environments, such as the Okumura–Hata model, the COST 231 model, and the Motley–Keenan model. Due to the specificity of the CBTC application, to the best of our knowledge, there is a channel model with leaky waveguide in CBTC systems. As the channel with leaky waveguide is linear as shown earlier, similar to the expression of the Okumura–Hata model [6], we propose a model of the CBTC channel with leaky waveguide as follows:

$$PL(d) = PL(0) + nd + X_{ss} \qquad (5.1)$$

where:

PL(0) is the difference between the power leaked from the beginning of the leaky waveguide and the input power

n is the path loss exponent whose unit is dB/m

d is the location of the receiving antenna relative to the beginning of the leaky waveguide

X_{ss} is the small-scale fading the is a random variable

PL(0) depends on the dimensions of the leaky waveguide and the slots. There are two key parameters, n and X_{ss}, in the model that will be determined in Section 5.4.2 and 5.4.3 respectively.

5.4.2 Determination of the Path Loss Exponent

According to the slopes of fitting lines, we can get the average value of the path loss exponent as follows:

$$n = \frac{1}{t} \sum_{m}^{t} n_m \qquad (5.2)$$

where:

t is the total number of measurements

n_m is the slope of the fitting line of the mth measurement

In our measurements, the average value of the path loss exponent is 0.0136 dB/m.

The consecutive slot radiation can be represented by separate successive equivalent magnetic dipoles fed by the same power [1]. The equivalent method can be effective to describe the fading tendency of the channel with leaky waveguide, but it may not be reasonable to assume that the magnetic dipoles are fed by the same power when the length of the leaky waveguide is so large, because the transmission loss should not be ignored. Then, we need to calculate the transmission loss as follows [1]:

$$\alpha = \frac{20}{\ln 10} \frac{1}{b^{3/2}} \sqrt{2\pi\varepsilon_0 \rho_{wg}} \frac{(f_c/f)^2 + b/2a(f/f_c)^2}{\left[1 - (f_c/f)^2\right]^{1/2}} \tag{5.3}$$

where:

ε_0 is the dielectric constant of vacuum (the value is 8.85×10^{-12} F / m)

ρ_{wg} is the resistivity of the material of leaky waveguide (the value is $2.9 \times 10^{-8} \Omega \cdot m$)

a and b are the dimensions of leaky waveguide

f_c is the cut-off frequency of the leaky waveguide

As each slot is taken as a magnetic dipole, the electric field radiated from a slot is as follows [7]:

$$E = E_0 \frac{\sin\theta}{r} e^{-j\beta r} \tag{5.4}$$

where:

E_0 is the electric field amplitude of the slot

E is the electric field strength radiated from the slot

θ is the intersection angle between the line from the receiving point to the slot and the plane of leaky waveguide

β is the propagation constant of free space

r is the distance between the slot and the receiving point

Therefore, the leaky waveguide can be taken as a magnetic dipole antenna array, and then the radiated electric field strength can be expressed as follows:

$$E = \sum_{k=1}^{N} \left| E_0 e^{j\alpha(k-1)d} \frac{\sin\theta_k}{r_k} e^{-j\beta r_k} \right| \tag{5.5}$$

where:

E_0 is the electric field amplitude of the first slot of leaky waveguide, normalized as 1 V/m

N is the total number of slots

α is the transmission loss of leaky waveguide (its unit is linear here, but not dB)

θ_k is the intersection angle between the line from the receiving point to the kth slot and the plane of leaky waveguide

r_k is the distance between the receiving point to the kth slot

In order to theoretically prove the linear model of the large-scale fading, the equivalent magnetic dipole method is applied as the supplement to the measurements. The simulation results are shown in Figure 5.4, which are approximately straight line and almost parallel to the fitting lines. We use MATLAB® to implement the simulation of the equivalent magnetic dipole method. The length of leaky waveguide

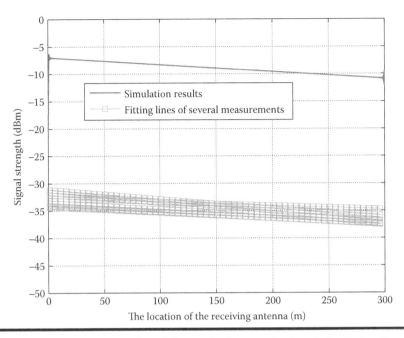

Figure 5.4 Simulation results of the equivalent method and the fitting lines of several measurements.

is 300 m, and the distance between adjacent slots is set as 60 mm. The receiving antenna is set as 320 mm right above the leaky waveguide, and it can receive electromagnetic waves leaked from slots based on Equation 5.5. The operation frequency is 2.412 GHz. Considering the transmission loss of leaky waveguide, the sum of electric strength from different slots is obtained. For ease of calculation, the electric field amplitude of the first slot is normalized. Moreover, in simulations, it is difficult to consider the insertion loss, the feeder line loss, and the splicing loss, which generally cannot be ignored in measurements. As a result, there is difference between the simulation results and the fitting lines. However, the simulation shows that the distribution of electric field strength is approximately linear and the slope is close to the transmission loss.

From the derivative of Equation 5.5 with respect to d, we can find that the path loss exponent is mostly dependent on the value of transmission loss α. The transmission loss of the leaky waveguide can be calculated according to Equation 5.3, which is 0.0139 dB/m denoted as n_t. Comparing the parameters n with n_t, we find that n_t is almost equal to n, which means that we can use the transmission loss as the path loss exponent of the radio channel with leaky waveguide. And it will be convenient and effective for practical engineering, especially the link budget.

In Equation 5.1, the path loss exponent n has been obtained, and PL(0) is an inherent parameter of one kind of leaky waveguide. Next, we will discuss the distribution model of small-scale fading.

5.4.3 Determination of the Small-Scale Fading

The statistical model of the small-scale fading needs to be determined to show how much the signal level can vary on the basis of the large-scale fading. A number of different distributions have been proposed for small-scale amplitude fading in indoor and outdoor environments, including Rice, Rayleigh, Nakagami, Weibull, and log-normal distributions. In order to determine the distribution of the small-scale fading of the channel with leaky waveguide, we first need to remove the effects of the large-scale fading. And the length of small-scale area (SSA) should be determined as 20λ to get enough data to estimate the model precisely for one SSA [8]. As the empirical decorrelation distance is $\lambda/2$, we select one sample from the measurement results in each $\lambda/2$ section. And the samples in different decorrelation distance areas are spatial independent (uncorrelated).

We select five kinds of classic distributions mentioned in the preceding text as the candidate models of the small-scale fading for the channel with leaky waveguide. These models are also used in [8], where the AICc is used. AICc is also called the second-order AIC, which is defined as follows [9]:

$$\text{AIC}_{i,j} = -2\sum_{n=1}^{N_j} \log_e(l(\hat{\theta}_{i,j} \mid x_{i,n})) + 2k_j$$

$$\text{AICc}_{i,j} = \text{AIC}_{i,j} + \frac{2k_j(k_j+1)}{N_i - k_j - 1}$$

(5.6)

where:

i means the ith SSA

j means the jth candidate model

$\hat{\theta}_{i,j}$ means the estimated parameters of the jth candidate model for the ith SSA using the maximum likelihood estimator (MLE)

$x_{i,n}$ is the nth sample of the ith SSA

k_j is the number of parameters of the jth candidate model

N_i is the total number of samples of the ith SSA

For each SSA, we calculate the value of AICc of each candidate model and select the best-fit model with the lowest value of AICc. Figure 5.5 shows that the log-normal distribution provides the best fit in a majority of the cases, about 54.7%. This is valid for 20 measurements in Beijing Subway Yizhuang Line. Therefore, we can see that the log-normal model is the best parametric fit to the distribution of the small-scale fading.

In addition, we show the plots of the empirical cumulative distribution functions (CDFs) of measurement data and the candidate models in Figure 5.6, where

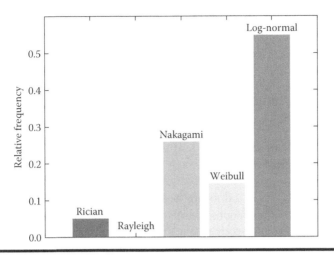

Figure 5.5   Relative frequencies of AICc selecting a candidate distribution as the best fit to the distribution of small-scale fading amplitudes.

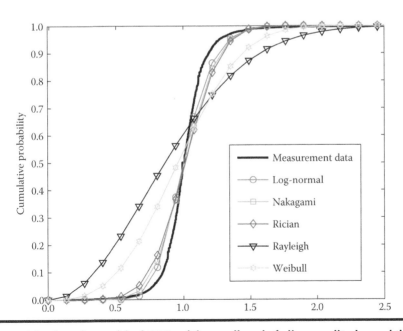

Figure 5.6   Sample empirical CDFs of the small-scale fading amplitudes and their theoretical model fits.

the *x*-axis is the fading amplitude whose unit is linear. The same observation can be obtained through CDFs. The log-normal distribution provides the closest fit to the measurement results. The expression of log-normal distribution is

$$p(x;\mu,\sigma) = \frac{1}{x\sigma\sqrt{2\pi}} e^{\left(-\ln x - \mu/2\sigma^2\right)}, x > 0 \tag{5.7}$$

where:

    μ and σ are the mean and standard deviations of the variable's natural logarithm, respectively

As we can obtain μ_{dB} and σ_{dB} through the MLE method at every SSA with the exact location information, the relationship between the parameters of log-normal distribution and the location of the receiving antenna can be built. We get the mean values of μ_{dB} and σ_{dB} of different measurements at the same SSA, as shown in Figure 5.7, where the *x*-axis is the distance from the beginning of the leaky waveguide to the midpoint of each SSA. We build the relationship between the parameters of log-normal distribution and the location of the receiver. The average of μ_{dB} is 0 and the average of σ_{dB} is almost −10. At the beginning and the end of the leaky waveguide, the value of σ_{dB} increases quickly, which means that the amplitudes fluctuate wildly. The reason is that the beginning of the leaky waveguide is out of the radiation coverage of the first

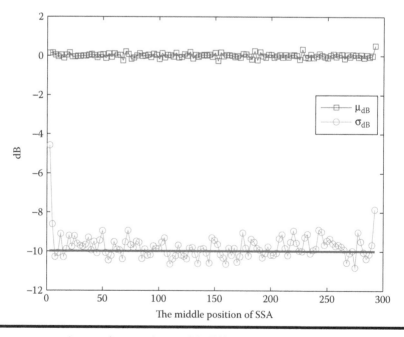

Figure 5.7 **Variance of μ_{dB} and σ_{dB} with different receiving points.**

slot. Similarly, the end of the leaky waveguide is not covered by the radiation of the last slot. In other words, at the beginning and the end of leaky waveguide, it is natural propagation that has higher attenuation and more severe fluctuations, which is the same as the simulation results in Figure 5.4. However, μ_{dB} is almost independent of the location of the receiving antenna, which shows the stability of leaky waveguide.

5.5 Conclusion

In the chapter, based on the measurements, we have proposed a linear path loss model for the large-scale fading of CBTC radio channel with leaky waveguide. The path loss exponent can be approximated by the transmission loss of leaky waveguide. In addition, we have shown that, with the AICc method, the small-scale fading follows log-normal distribution. The parameters of log-normal distribution have been determined.

References

1. M. Heddebaut. Leaky waveguide for train-to-wayside communication-based train control. *IEEE Trans. Veh. Tech.*, 58(3):1068–1076, 2009.
2. L. Zhu, F. R. Yu, B. Ning, and T. Tang. Cross-layer handoff design in MIMO-enabled WLANs for communication-based train control (CBTC) systems. *IEEE J. Sel. Areas Commun.*, 30(4):719–728, 2012.
3. L. Zhu, H. Wang, and B. Ning. An experimental study of rectangular leaky waveguide in CBTC. In *Proceedings of the IEEE Intelligent Vehicles Symposium*, pp. 951–954, 2009.
4. Q. Wang, T. Tayamachi, I. Kimura, and J. Wang. An on-body channel model for UWB body area communications for various postures. *IEEE Trans. Antennas Propagation*, 57(4):991–998, 2009.
5. A. Fort, J. Ryckaert, C. Desset, P. De Doncker, P. Wambacq, and L. Van Biesen. Ultra-wideband channel model for communication around the human body. *IEEE J. Sel. Areas Commun.*, 24(4):927–933, 2006.
6. A. F. Molisch. *Wireless Communications*. John Wiley & Sons, New York, 2011.
7. John D. Kraus and Ronald J. Marhefka. *Antennas: For All Applications*. Publishing House of Electronics Industry, Beijing, China, 2008.
8. S. Wyne, A. P. Singh, F. Tufvesson, and A. F. Molisch. A statistical model for indoor office wireless sensor channels. *IEEE Trans. Wireless Commun.*, 8(8):4154–4164, New York, 2009.
9. K. P. Burnham and D. R. Anderson. *Model Selection and Multimodel Inference: A Practical Information-Theoretic Approach*. Springer, 2002.

Chapter 6

Communication Availability in Communications-Based Train Control Systems

Li Zhu and F. Richard Yu

Contents

6.1 Introduction 94
6.2 Related Works 95
6.3 Proposed Data Communication Systems with Redundancy 95
    6.3.1 Overview of CBTC and Data Communication System 96
    6.3.2 Proposed Data Communication Systems with Redundancy 97
6.4 Data Communication System Availability Analysis 99
6.5 Modeling Data Communication System Behavior with DSPNs 103
    6.5.1 Introduction to DSPNs 103
    6.5.2 DSPN Formulation 104
    6.5.3 DSPN Model Solutions 105
6.6 Numerical Results and Discussions 110
    6.6.1 System Parameters 110
    6.6.2 Model Soundness 111
    6.6.3 Availability Improvement 111
6.7 Conclusion 113
References 113

6.1 Introduction

A data communication system, which is the basis for train control, is one of the key subsystems for communications-based train control (CBTC). A couple of wireless communication technologies have been adopted in CBTC, such as Global System for Mobile Communications-Railway (GSM-R) and wireless local area network (WLAN). For urban mass transit systems, WLAN is commonly used due to the available commercial off-the-shelf equipment, open standards, and interoperability [1–3].

CBTC systems have stringent requirements for communication availability [2]. Whereas in commercial wireless networks, less service availability means less revenues and/or poor quality of services [4], in CBTC systems, it could cause train derailment, collision, or even catastrophic loss of life or assets [5]. Therefore, it is important to ensure the train–ground communication availability in CBTC systems.

There are several WLAN-based CBTC systems deployed around the world, such as Las Vegas Monorail from Alcatel [6] and Beijing Metro Line 10 from Siemens [7]. Most system integrators claim that redundancy is used in their systems. However, they do not reveal the details about redundancy due to confidential considerations. Moreover, availability analysis is largely ignored in the literature of CBTC systems.

In this chapter, we study the availability issue of WLAN-based data communication systems in CBTC. The contributions of this chapter are as follows:

1. We propose two WLAN-based data communication systems with redundancy to improve the availability in CBTC systems.
2. The availability of WLAN-based data communication systems is analyzed using continuous-time Markov chain (CTMC) model [8], which has been successfully used in call admission control (CAC) in mobile cellular networks [9] and wireless channel modeling [10], among others. The transmission errors due to dynamic wireless channel fading and handoffs that take place when the train crosses the border of two successive access points (APs)'s coverage areas are considered as the main causes of system failures.
3. We model the WLAN-based data communication system behavior using deterministic and stochastic Petri net (DSPN) [11], which is a high-level description language for formally specifying complex systems. The DSPN solution is used to show the soundness of our proposed CTMC model. DSPN provides an intuitive and efficient way of describing the complex system behavior and facilitates the modeling of system steady-state probability.
4. Using numerical examples, we compare the availability of the two proposed WLAN-based data communication systems with an existing system that has no redundancy. The results show that the proposed data communication systems with redundancy have much higher availability than the existing system.

The rest of this chapter is organized as follows: Section 6.2 introduces the related works. Section 6.3 presents an overview of CBTC and the proposed data communication systems with redundancy. Section 6.4 discusses the CTMC model, with its state space for each configuration. Section 6.5 describes the DSPN modeling approach. Section 6.6 presents numerical results. Finally, Section 6.7 concludes the chapter.

6.2 Related Works

Documented research has investigated the availability issue of commercial wireless networks. The authors of [12] analyze the availability and reliability of wireless multi-hop networks with stochastic link failures. They provide a method to forecast how the introduction of redundant nodes increases the reliability and availability of such networks. A novel wireless communication infrastructure is presented in [13] for emergency management. Several schemes are proposed in the infrastructure to improve the system reliability. In [14], techniques for composite performance and availability analysis are discussed in detail through a queuing system in a wireless communication network. Three modeling approaches are illustrated for composite performance and availability analysis in their work. In [15,16], different WLAN topologies are considered in home WLANs, and the simulation results show that redundancy can greatly improve the system availability of home WLANs.

There are also some works about availability in CBTC data communication systems. For trunk railway systems, the authors of [17] study the GSM-R systems with stochastic Petri net (SPN) model in European train control systems (ETCSs). A similar model for ETCS with a unified modeling language (UML) software tool, STOCHARTS, is proposed in [18]. Based on the parameters taken from the ETCS standards, the results show that the communication requirement in ETCS is very strict for train control systems. For urban mass transit systems, the authors of [5] study the CBTC data communication system with simple redundancy, and the reliability and availability are analyzed. However, the redundancy schemes in their study are exploratory. They do not consider the impact of AP deployment space on system availability performance, and the analysis approach may not be comprehensive.

6.3 Proposed Data Communication Systems with Redundancy

In this section, we first present an overview of CBTC and its basic configuration of data communication system based on WLANs. Then, the proposed data communication systems with redundancy are presented.

6.3.1 Overview of CBTC and Data Communication System

A simple view of a CBTC system is illustrated in Figure 6.1. In this system, continuous bidirectional wireless communications between each station adapter (SA) on the train and the wayside AP are adopted instead of the traditional fixed-block track circuit. The railway line is usually divided into areas. Each area is under the control of a zone controller (ZC) and has its own wireless transmission system. The identity, location, direction, and speed of each train are transmitted to the ZC. The wireless link between each train and the ZC must be continuous to ensure that the ZC knows the locations of all the trains in its area at all times. The ZC transmits to each train the location of the train in front of it, and a braking curve is given to enable it to stop before it reaches that train as well. Theoretically, two successive trains can travel together as close as a few meters in between them, as long as they are traveling at the same speed and have the same braking capability.

Data communication systems are primarily designed to connect each component of CBTC systems: ZCs, APs along a railway, and train aboard equipments. A basic configuration of WLAN-based data communication system is shown in Figure 6.1. Following the philosophy of open standards and interoperability [2], the backbone network of the data communication system, which mainly includes Ethernet switches and fiber-optic cabling, is based on the IEEE 802.3 standard. The wireless portions of the data communication system, which consist of APs along the railway and SAs on the train, are based on the IEEE 802.11.

In contrast to the backbone network, the wireless links are more prone to channel degradation in railway environments. The low reliability of the wireless portion of the data communication system is mainly caused by the following:

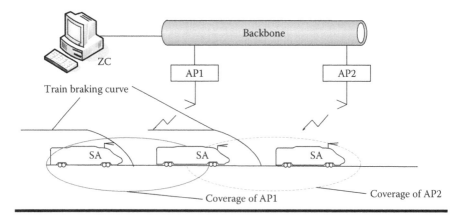

Figure 6.1   CBTC system.

1. Transmission errors due to dynamic channel fading in railway environments.
2. Handoffs that take place every time the train crosses the border of two successive APs's coverage areas. The communication link will be lost for a short period of time during the handoff process.

In order to improve the communication availability in CBTC systems, we propose two data communication systems with redundancy, which will be presented in Section 6.3.2.

6.3.2 Proposed Data Communication Systems with Redundancy

In order to describe our system more easily, we first define two different kinds of links for the data communication system.

1. Active link: A link that is currently used between an SA and its associated AP
2. Backup link: A link that is currently not used between an SA and its associated AP, but can be used in case of failure of the active link

If the SA is in the coverage of more than one AP, the active link is the one that the SA associates with the AP, which provides the better signal-to-noise ratio (SNR). In case of failure of the active link, the SA can associate with another AP. We call this link a backup link.

The data communication system with basic configuration is shown in Figure 6.1. In this system, only one AP with directional antenna is deployed in each location. The head directional antenna of the train is connected to the SA. There is only one active link for the train at any time. No backup link exists in the basic configuration.

We propose two data communication systems with redundancy configurations. The first proposed system is shown in Figure 6.2. In this system, two APs each with one directional antenna (facing in opposite directions) are used in each location, which are, respectively, connected to two backbone networks (i.e., Backbone network 1 and Backbone network 2). Two directional antennas (i.e., head antenna and tail antenna) are connected with two independent SAs on the train. The two APs in each location have different service set identifiers (SSIDs), and the two SAs on the train also have different SSIDs. Normally, SAs can only associate with the AP with the same SSID. No backup link exists in this redundancy configuration. Instead, there are two active links between the train and the ground. The communication will not be interrupted if only one of the active links fails because of deep channel fading or handoff. We assume that there is no chance that handoffs happen at both SAs at the same time. This assumption is reasonable in practice, because the APs can be appropriately deployed so

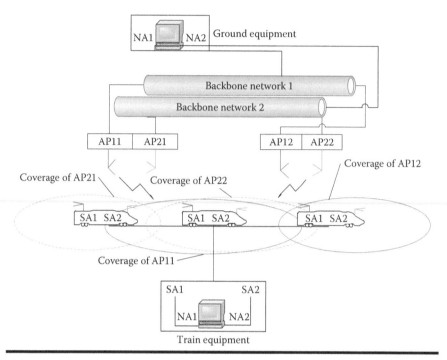

Figure 6.2   First proposed data communication system with redundancy and no backup link.

that when one SA on the train is at the coverage edge, the other one is still in the coverage of another AP. We name this system as the data communication system with redundancy and no backup link.

The second proposed system with redundancy configuration is shown in Figure 6.3. In this system, only one AP with directional antenna is used in each location. Two directional antennas, i.e., head antenna and tail antenna, are connected with independent SA at each end of the train. All the SAs and APs in the system have the same SSID. Compared to the first proposed system, the AP space in the second system is halved to make sure that any SA on the train is on the coverage of two APs. There is one active link and backup link for both the head SA and the tail SA. The corresponding backup link will become active when the active link fails. The communication will be interrupted only if all four links fail. We name this system as the data communication system with redundancy and backup link.

We take the data communication system with redundancy and no backup link (as shown in Figure 6.2) as an example to illustrate how to implement it in practice. There are normally two active links between the train and the ground. In real systems, we connect each end of the active link with a network adapter.

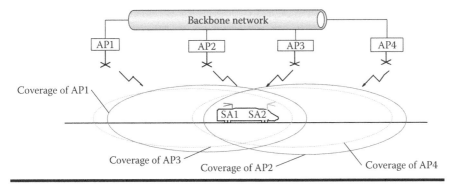

Figure 6.3 Second proposed data communication system with redundancy and backup link.

As shown in Figure 6.2, there are two network adapters for both the train and the ground equipment. A high-level protocol needs to be designed between the train and the ground to keep the data from one active link and delete the redundant data from the other active link. Communication will not be interrupted if only one of the active links fails. The availability of these systems will be analyzed in Section 6.4.

6.4 Data Communication System Availability Analysis

System availability is the probability that the system is operating at a specified time [19]. It is primarily used in repairable systems, where brief interruptions in service can be tolerated.

The WLAN-based data communication system is modeled as a CTMC [8]. We define the link state as $s = (i, j, k)$, where i is the number of active links, j is the number of failed links caused by deep channel fading, and k is the number of failed links caused by handoff.

According to CTMC theory [8], let $P_s(t)$ be the unconditional probability of the CTMC being in state s at time t. Then the row vector,

$$P(t) = [P_1(t), P_2(t), ..., P_s(t)] \qquad (6.1)$$

represents the transient state probability vector of the CTMC. Given $P(0)$, the behavior of the CTMC can be described by the following Kolmogorov differential equation:

$$\frac{\mathrm{d}}{\mathrm{d}t} P(t) = P(t) \times Q \qquad (6.2)$$

where:

$P(0)$ represents the initial probability vector (at time $t = 0$) of the CTMC

Q is the infinitesimal generator matrix

$Q = [q_{ss'}]$ represents the transition rate from state s to state s'

The diagonal elements are $q_{ss} = -\sum_{s' \neq s} q_{ss'}$

The CTMC model for data communication system with basic configuration is shown in Figure 6.4. In this model, state $(1,0,0)$ is the only service state where the only existing link is active, $\lambda 1$ is the deep channel fading rate, $\lambda 2$ is the handoff rate, $\mu 1$ is the fading recovery rate, and $\mu 2$ is the handoff recovery rate.

The infinitesimal generator matrix for this model is

$$Q1 = \begin{pmatrix} -(\lambda 1 + \lambda 2) & \lambda 1 & \lambda 2 \\ \mu 1 & -\mu 1 & 0 \\ \mu 2 & 0 & -\mu 2 \end{pmatrix} \qquad (6.3)$$

With the same link state $s = (i, j, k)$, the CTMC model for the first proposed data communication system with redundancy and no backup link is shown in Figure 6.5.

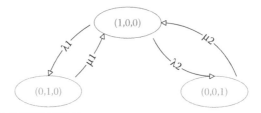

Figure 6.4 CTMC model for the data communication system with basic configuration.

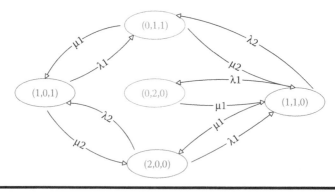

Figure 6.5 CTMC model for the first proposed data communication system with redundancy and no backup link.

In this model, (2,0,0), (1,0,1), and (1,1,0) are service states where at least one link is active. As explained in Section 6.2, state (0,0,2) does not exist because the head SA and the tail SA will not handoff at the same time. States (0,1,1) and (0,2,0) are out-of-service states for lack of active links.

The infinitesimal generator matrix for this model is

$$
Q2 = \begin{pmatrix}
-(\lambda 1+\lambda 2) & \lambda 1 & \lambda 2 & 0 & 0 \\
\mu 2 & -(\lambda 1+\lambda 2) & 0 & \lambda 1 & 0 \\
\mu 1 & 0 & -(\lambda 1+\lambda 2+\mu 1) & \lambda 2 & \lambda 1 \\
0 & \mu 1 & \mu 2 & -(\mu 1+\mu 2) & 0 \\
0 & 0 & \mu 1 & 0 & -\mu 1
\end{pmatrix} \quad (6.4)
$$

Next, we extend the link state $s=(i,j,k)$ to $\overline{s}=(h,i,j,k)$, where h is the total number of active links and backup links, $\mu 3$ is the rate of successful transitions from backup links to active links. The CTMC model for the second proposed data communication system with redundancy and backup link is shown in Figure 6.6. In this model, (2,0,1,1), (1,0,2,1), and (0,0,4,0) are out-of-service states where no active link exists. All the other states are service states that have at least one active link.

The infinitesimal generator matrix for this model is

$$
Q3 = \begin{pmatrix}
A_1 & \lambda 1 & 0 & 0 & 0 & 0 & \lambda 2 & 0 & 0 & 0 & 0 & 0 & 0 \\
\mu 1 & A_2 & \mu 3 & 0 & 0 & 0 & 0 & 0 & 0 & 0 & \lambda 2 & 0 & 0 \\
\mu 1 & 0 & A_3 & \lambda 1 & 0 & 0 & 0 & \lambda 2 & 0 & 0 & 0 & 0 & 0 \\
0 & 0 & \mu 1 & A_4 & \mu 3 & 0 & 0 & 0 & 0 & 0 & 0 & \lambda 1 & 0 \\
 & & & & \cdots & \cdots & & & & & & & \\
 & & & & \cdots & \cdots & & & & & & & \\
 & & & & \cdots & \cdots & & & & & & & \\
0 & 0 & 0 & 0 & 0 & \mu 2 & 0 & 0 & \mu 1 & A_5 & 0 & 0 & 0 \\
0 & \mu 2 & 0 & 0 & 0 & 0 & \mu 1 & 0 & 0 & 0 & A_6 & 0 & 0 \\
0 & 0 & 0 & \mu 2 & 0 & 0 & 0 & \mu 1 & 0 & 0 & 0 & A_7 & 0 \\
0 & 0 & 0 & 0 & 0 & \mu 1 & 0 & 0 & 0 & 0 & 0 & 0 & -\mu 1
\end{pmatrix} \quad (6.5)
$$

where:

$A_1 = -(\lambda 1+\lambda 2)$

$A_2 = -(\mu 1+\mu 3+\lambda 2)$

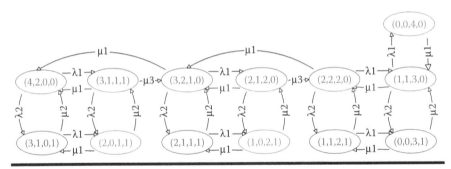

Figure 6.6 **CTMC model for the first proposed data communication system with redundancy and backup link.**

$$A_3 = -(\lambda 1 + \lambda 2 + \mu 1)$$
$$A_4 = -(\mu 1 + \mu 3 + \lambda 2)$$
$$A_5 = -(\mu 1 + \mu 2)$$
$$A_6 = -(\mu 1 + \mu 2)$$
$$A_7 = -(\mu 1 + \mu 2)$$

In order to derive the system availability, let π_n be the steady-state probability of state n for the three CTMCs. π_n shall satisfy the following equations [8]:

$$(\pi_1, \pi_2, ..., \pi_N) \times Q = (0, 0, ..., 0) \tag{6.6}$$

$$\sum_{n=1}^{N} \pi_n = 1 \tag{6.7}$$

where n is the number of states for the three CTMCs.

The steady-state probability can be obtained by resolving the previous two equations, and the system availability for each of the three systems can be derived from the steady-state probability as follows:

$$A = \sum_{w \in W} \pi_w \tag{6.8}$$

where W is the aggregation of service states.

The system unavailability UA is then derived as

$$UA = 1 - A \tag{6.9}$$

6.5 Modeling Data Communication System Behavior with DSPNs

The main problem in our proposed CTMC model in Section 6.3 is that we assume that the time between two successive handoffs is exponentially distributed. However, the distance between successive APs and the train speed both may not follow exponential distribution. In order to show the soundness of the approximation in our CTMC model, we formulate the data communication system behavior with DSPNs.

The motivations of selecting the DSPN approach are as follows: (1) As a kind of SPN, DSPN provides an intuitive and efficient way of describing the link failure behaviors in WLAN-based data communication systems, especially facilitating handoff behavior modeling. (2) DSPN allows timed transitions to have an exponentially distributed time delay or a deterministic timed delay, which can accurately model the situation when the time between two successive handoffs is relatively constant. (3) SPN has been successfully used to analyze system availability in safety-critical on-demand systems [20] and industry plants [21], among others.

6.5.1 Introduction to DSPNs

A Petri net is a directed bipartite graph with two types of nodes called places and transitions, which are represented by circles and rectangles (or bars), respectively [22]. Arcs connecting places to transitions are referred to as input arcs, whereas the connections from transitions to places are called output arcs. A nonnegative integer (the default value is 1) may be associated with an arc, which is referred to as multiplicity or weight. Places correspond to state variables of the system, whereas transitions correspond to actions that induce changes of states. A place may contain tokens that are represented by dots in the Petri net. The state of the Petri net is defined by its marking, which is represented by a vector $M = (l_1, l_2, ..., l_k)$, where $l_k = M(p_k)$ is the number of tokens in place p_k. Here, $M(\cdot)$ is a mapping function from a place to the number of tokens assigned to it. A transition is enabled if the number of tokens in each of its input places is larger than the weight of its corresponding input arc. An enabled transition can fire, and as many tokens as the weight of the corresponding input arcs are moved from the input place to the output place.

SPNs are one kind of Petri nets in which an exponentially distributed time delay is associated with each transition [22]. Generalized SPNs (GSPNs) extend the modeling power of SPN and divide the transitions into two classes: the exponentially distributed timed transitions (represented by blank rectangles), which are used to model the random delays associated with the execution of activities, and immediate transitions (represented by bars), which are devoted to the representation of logical actions that do not consume time. DSPN [23] further extends GSPN in that it allows timed transitions to have an exponentially distributed time delay or a deterministic timed delay (represented by filled rectangles). The firing rate of the timed transitions may be marking dependent.

6.5.2 DSPN Formulation

In this section, we model the data communication systems with different configurations using different DSPNs.

For the data communication system with basic configuration, the corresponding DSPN is shown in Figure 6.7. Place P_{active} indicates the number of active link, and it initially has one token. The system is on service when the only token is in P_{active}. The tokens in place P_{fading} indicate the number of failed links caused by deep channel fading, and the tokens in place $P_{handoff}$ indicate the number of failed links caused by handoff. T_{fading} is an exponentially distributed timed transition, which denotes a fading process. When it fires, a token will move from place P_{active} to P_{fading}. $T_{fading\_recovery}$ is the corresponding exponentially distributed timed transition that denotes a fading recovery process. When it fires, a token will move from place P_{fading} to P_{active}. Transition $T_{handoff}$ is a deterministic timed transition for service process. When it fires, a token will move from place P_{active} to $P_{handoff}$. $T_{handoff\_recovery}$ is the corresponding exponentially distributed timed transition that denotes a handoff recovery process. When it fires, a token will move from place $P_{handoff}$ to P_{active}.

Similarly, the DSPN model for the data communication systems with redundancy configurations is shown in Figure 6.8. Compared with the DSPN for basic configuration, in this DSPN model, place P_{active} initially has two tokens. The system is on service only if P_{active} has one token inside. The explanation of other places and transitions in the model is the same as that of the DSPN model for basic configuration.

For the data communication systems with redundancy and backup link configurations, the corresponding DSPN is shown in Figure 6.9. A place P_{backup} is added into this DSPN model compared to other two models. The tokens in P_{backup} indicate the number of backup link, and there are initially two tokens in P_{backup}

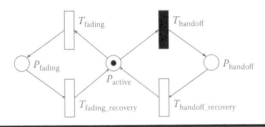

Figure 6.7 DSPN model for the data communication system with basic configuration.

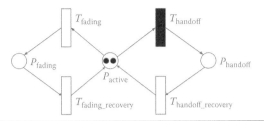

Figure 6.8 DSPN model for the data communication system with redundancy and no backup link.

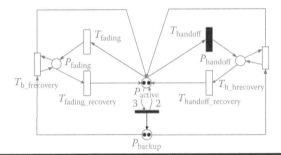

Figure 6.9 DSPN model for the proposed data communication system with redundancy and backup link.

which correspond to two active links. When an active link fails due to deep channel fading or handoff, a back link will become active after an exponentially distributed time. This process is denoted by two transitions $T_{b\_frecovery}$ and $T_{b\_hrecovery}$. The explanation of other places and transitions in the model is the same as that of the other two DSPN models presented earlier.

6.5.3 DSPN Model Solutions

As shown in Figures 6.7 through 6.9, the number of deterministic transitions for each marking is at most one in our DSPN models. According to [24], if at most one deterministic transition is allowed to be enabled in each marking, the state probabilities of a DSPN can be obtained analytically rather than by simulation.

The stochastic process $\{D(t), t > 0\}$ underlying the DSPN is formed by the variations of the markings over the time domain. $\{D(t)\}$ is the marking of the DSPN at time t. According to [24], when at most one deterministic transition is enabled in each marking of the DSPN, the marking process $\{D(t), t > 0\}$ is a Markov regenerative process (MRGP). We include the proof for completeness next.

Let Ω be the set of all markings of a DSPN. It is also the state space of the marking process of the DSPN. For example, $\Omega = \{m1, m2, m3\}$ is the state space of the marking process of the DSPN in Figure 6.7, where $m1$, $m2$, and $m3$ indicate that the only token in the model is in P_{fading}, P_{active}, and P_{handoff}, respectively. Consider a sequence $\{T_n, n > 0\}$ of epochs when the DSPN is observed. Let $T_0 = 0$ and define $\{T_n, n > 0\}$ recursively as follows:

Suppose $D(T_n) = m$,

1. If no deterministic transition is enabled in state m, define T_{n+1} to be the first time after T_n that a state change occurs. If no such time exists, we set $T_{n+1} = \infty$.
2. If a deterministic transition is enabled in state m, define T_{n+1} to be the time when the deterministic transition fires or is disabled.

With the above definition of $\{T_n, n > 0\}$, let $Y_n = D(T_n)$. We first show that $\{(Y_n, T_n), n \geq 0\}$ embedded in the marking process of DSPN is a Markov renewal sequence, that is,

$$
\begin{aligned}
P\{Y_{n+1} = j, T_{n+1} - T_n \leq t \,|\, Y_n = i, T_n, Y_{n-1}, T_{n-1}, \ldots, Y_0, T_0\} \\
= P\{Y_{n+1} = j, T_{n+1} - T_n \leq t \,|\, Y_n = i\} = P\{Y_1 = j, T_1 \leq t \,|\, Y_0 = i\}
\end{aligned}
\tag{6.10}
$$

The above equation can be explained in the following two cases:

1. No deterministic transition is enabled, that is, all the transitions that are enabled in state i at time T_n are exponential transitions. In this case, due to the memoryless property of the exponential random variable, the future of the marking process depends only on the current state i and does not depend upon the past history or the time index n.
2. Exactly one deterministic transition is enabled in state i. There may be other exponential transitions enabled in state i. In this case, T_{n+1} is the next time when the deterministic transition fires or is disabled. The joint distribution of Y_{n+1} and $(T_{n+1} - T_n)$ will depend only on the state at time T_n. It is thus independent of the past history and the time index n.

Now considering the marking process after time T_n, namely, $\{D(T_n + t), t > 0\}$. Given the history $\{D(u), 0 < u < T_n, D(T_n) = i\}$, we can see from Equation 6.10 that the stochastic behavior of $\{D(T_n + t) > 0, t \geq 0\}$ depends only on $D(T_n) = i$. Therefore,

$$
\begin{aligned}
\{D(T_n + t) > 0, t \geq 0 \,|\, D(u), 0 < u < T_n, D(T_n) = i\} \\
\doteq \{D(T_n + t) > 0, t \geq 0 \,|\, D(T_n) = i\} \\
\doteq \{D(t) > 0, t \geq 0 \,|\, D(0) = i\}
\end{aligned}
\tag{6.11}
$$

where $\doteq$ denotes equality in distribution. Equation 6.11 proves that $\{D(t), t > 0\}$ is a Markov regenerative process.

Let $t \to \infty$. We can see $\{Y_n, n \geq 0\}$ is a discrete time Markov chain with transition probability matrix $K(\infty)$. It is also called the embedded Markov chain (EMC) for the DSPN. Matrix $K(t) = [K_{ij}(t)]$ is called the kernel and $K_{ij}(t)$ satisfies

$$K_{ij}(t) = P\{Y_1 = j, T_1 \leq t \mid Y_0 = i\} \tag{6.12}$$

Once we prove that the marking process under a DSPN is an MRP, the distribution $p = (p_j)$ of the steady-state probabilities of the MRGP, which is also the fraction of time the marking process spends in state j, can be calculated by the following steps:

1. Calculate the kernel $K(t) = [K_{m,n}(t)]_{m,n}(m, n \in \Omega)$ of the marking process of DSPN as follows:
   a. For state m when $\Gamma(m) = \Phi$,

$$K_{m,n}(t) = \begin{cases} 0 & \Lambda_m = 0 \\ \dfrac{\lambda(m,n)}{\Lambda_m}\left(1 - e^{-\Lambda_m t}\right) & \Lambda_m > 0 \end{cases} \tag{6.13}$$

where:

$\Gamma(m)$ is the set of deterministic transitions enabled in state m

$\Gamma(m) = \Phi$ means that no deterministic transition is enabled in m

$\lambda(m,n)$ is the transition rate from marking m to n

$\Lambda_m = \sum_{n \in \Omega} \lambda(m, n)$

   b. For state m when $\Gamma(m) = d$ with τ being the firing time of transition d, if $n \in \Omega_\Upsilon(m)$ but $n \notin \Omega_\Gamma(m)$:

$$K_{m,n}(t) = \begin{cases} \left[e^{Q(m)t}\right]_{m,n} & t < \tau \\ \left[e^{Q(m)\tau}\right]_{m,n} & t \geq \tau \end{cases} \tag{6.14}$$

if $n \in \Omega_\Gamma(m)$ but $n \notin \Omega_\Upsilon(m)$:

$$K_{m,n}(t) = \begin{cases} 0 & t < \tau \\ \sum_{m' \in \Omega(m)} \left[e^{Q(m)\tau}\right]_{m,m'} \Delta(m', n) & t \geq \tau \end{cases} \tag{6.15}$$

if $n \in \Omega_\Gamma(m)$ but $n \in \Omega_\Upsilon(m)$:

$$K_{m,n}(t) = \begin{cases} 0 & t < \tau \\ \left[e^{Q(m)\tau}\right]_{m,n} + \sum_{m' \in \Omega(m)} \left[e^{Q(m)\tau}\right]_{m,m'} \Delta(m', n) & t \geq \tau \end{cases} \tag{6.16}$$

if $n \notin \Omega_\Gamma(m)$ but $n \notin \Omega_\Upsilon(m)$:

$$K_{m,n}(t) = 0 \tag{6.17}$$

In the above equations, $\Omega_\Upsilon(m)$ is the set of states that can be reached from m by firing a competitive exponential transition and $\Omega_\Gamma(m)$ is the set of states that can be reached by firing the deterministic transition. $Q(m)$ is the infinite generator matrix, which is formed as follows: for any $n \in \Omega$, the rate from n to $n' \in \Omega$ is given by $\lambda(n,n')$, and if $n \notin \Omega(m)$, the rates out of a marking n are zeros. $\Delta = [\Delta(n,n')]$ is the branching probability matrix, where $\Delta(n,n') = P\{\text{next marking is } n' \mid \text{current marking is } n \text{ and the deterministic transition fires}\}$.

2. Calculate the local kernel $E(t) = [E_{m,n}(t)]_{m,n} \ (m,n \in \Omega)$, where $E_{m,n}(t) = P\{D(t) = n, T_1 > t \mid Y_0 = m\}$, as follows:

   a. For state m when $\Gamma(m) = \Phi$,

   $$E_{m,n}(t) = \delta_{m,n} e^{-\Lambda_m t} \tag{6.18}$$

   where $\delta_{m,n}$ is the Kronecker function defined by $\delta_{m,n} = 1$ if $m = n$ and 0 otherwise. $\Gamma(m)$ is the set of deterministic transitions enabled in state m, and $\Gamma(m) = \Phi$ means that no deterministic transition is enabled in m. $\lambda(m,n)$ is the transition rate from marking m to n, and $\Lambda_m = \sum_{n \in \Omega} \lambda(m,n)$.

   b. For state m when $\Gamma(m) = d$ with τ being the firing time of transition d, if $n \in \Omega(m)$,

   $$E_{m,n}(t) = \begin{cases} \left[e^{Q(m)t} \right]_{m,n} & t < \tau \\ 0 & t \geq \tau \end{cases} \tag{6.19}$$

   if $n \notin \Omega(m)$,

   $$E_{m,n}(t) = 0 \tag{6.20}$$

3. Calculate the one-step transition probability matrix $P = [P_{m,n}]_{m,n} \ (m,n \in \Omega)$ for the EMC by the fact $P = K(\infty)$. Therefore, we get $P_{m,n}$ as follows:

   a. For state m when $\Gamma(m) = \Phi$,

   $$P_{m,n}(t) = \begin{cases} 0 & \Lambda_m = 0 \\ \dfrac{\lambda(m,n)}{\Lambda_m} & \Lambda_m > 0 \end{cases} \tag{6.21}$$

   b. For state m when $\Gamma(m) = d$ with τ being the firing time of transition d, if $n \in \Omega_\Upsilon(m)$ but $n \notin \Omega_\Gamma(m)$:

   $$P_{m,n}(t) = \left[e^{Q(m)\tau} \right]_{m,n} \tag{6.22}$$

if $n \in \Omega_{\Gamma}(m)$ but $n \notin \Omega_{\Upsilon}(m)$:

$$P_{m,n}(t) = \sum_{m' \in \Omega(m)} \left[e^{Q(m)\tau} \right]_{m,m'} \Delta(m', n) \tag{6.23}$$

if $n \in \Omega_{\Gamma}(m)$ but $n \in \Omega_{\Upsilon}(m)$:

$$P_{m,n}(t) = \left[e^{Q(m)\tau} \right]_{m,n} + \sum_{m' \in \Omega(m)} \left[e^{Q(m)\tau} \right]_{m,m'} \Delta(m', n) \tag{6.24}$$

if $n \notin \Omega_{\Gamma}(m)$ but $n \notin \Omega_{\Upsilon}(m)$:

$$P_{m,n}(t) = 0 \tag{6.25}$$

4. Calculate the steady-state probability $\xi = (\xi_j)$ for EMC from the following equation:

$$\xi = P \cdot \xi, \xi_j = 1 \tag{6.26}$$

5. Based on the kernel $K(t) = [K_{m,n}(t)]_{m,n}$ obtained in step 1, calculate the average time between epochs $\mu_m = E\{T_1 \mid Y_0 = m\}$ as,
   a. For state m when $\Gamma(m) = \Phi$,

$$\mu_m = E\{T_1 \mid Y_0 = m\} = \frac{1}{\Lambda_m} \tag{6.27}$$

   b. For state m when $\Gamma(m) = d$ with τ being the firing time of transition d,

$$\mu_m = E\{T_1 \mid Y_0 = m\} = \sum_{n \in \Omega} \int_0^{\tau} \left[e^{Q(m)t} \right]_{m,n} dt \tag{6.28}$$

6. Based on the local kernel $E(t) = [E_{m,n}(t)]_{m,n}$ obtained in step 2, calculate the expectation $\alpha_{mn} = E[\text{time spent by the marking process in state } n \text{ during } (0, T_1) \mid Y0 = m]$ as,

$$\alpha_{mn} = \int_0^{\infty} P\{D(t) = n, T_1 > t \mid Y_0 = m\} dt = \int_0^{\infty} E_{m,n}(t) dt \tag{6.29}$$

7. Based on the parameters $\xi = (\xi_j)$, μ_m, and α_{mn}, the distribution $p = (p_j)$ of the steady-state probabilities of the MRGP, which is also the fraction of time the marking process spends in state j, is given by [25]

$$p_j = \lim_{t \to \infty} P\{D(t) \mid Y_0 = m\} = \frac{\sum_{k \in \Omega} \xi_k \alpha_{kj}}{\sum_{k \in \Omega} \xi_k \mu_k} \qquad (6.30)$$

Once we get the steady-state probabilities of the MRGP $p = (p_j)$, the system availability can be calculated by adding the probability of the states or markings where the tokens in place P_{active} are not zero.

6.6 Numerical Results and Discussions

In this section, we first present the numerical parameters. Numerical results are presented to show the soundness of our availability models. Finally, the availability improvement of the proposed systems is presented.

6.6.1 System Parameters

In CBTC systems, APs are usually deployed to make their coverage areas overlap with each other in order to decrease the shadow zone. In our numerical examples, the distance between two successive APs is $l = 200\,\mathrm{m}$ for the first proposed data communication system with redundancy and no backup link, then the distance is $l = 100\,\mathrm{m}$ for the second proposed data communication system with redundancy and backup link. The handoff rate $\lambda 2$ is determined by the train velocity. Given the distance l between two successive APs and the train velocity v, the average time between two successive handoffs is l/v, which gives $(1/\lambda 2) = (l/v)$.

Similarly, the handoff end rate $\mu 2$ is determined by the handoff time. Given the average handoff time T_h, it can be calculated as $\mu 2 = l/T_h$. Other parameters we used for the CTMC model are shown in Table 6.1.

For the DSPN model parameters, according to [22], the average transition time for an exponentially distributed transition is the reciprocal of the transition rate. Therefore, all the exponentially distributed transition times in our DSPN models

Table 6.1 Parameters Used in Numerical Examples

Notation	Definition	Value
$\lambda 1$	Channel fading rate	$0.01\mathrm{s}^{-1}$
$\lambda 2$	Handoff rate	Determined by train velocity
$\mu 1$	Channel fading recovery rate	$0.2\mathrm{s}^{-1}$
$\mu 3$	Backup to active rate	$0.2\mathrm{s}^{-1}$

such as T_{fading} and $T_{\text{fading\_recovery}}$ can be calculated from the parameters in the CTMC models described above. And for the deterministic transitions T_{handoff}, as it indicates the average time between two successive handoffs, it can be calculated as $T_{\text{handoff}} = l/v$.

6.6.2 Model Soundness

Given all the parameters of the CTMC and DSPN models, Figure 6.10 illustrates the unavailability in the two proposed models for different configurations. As we can observe from the figure, the unavailability difference between the two models changes with model parameters, and the difference is not significant in most cases. Compared to the DSPN model, the unavailability error in the CTMC model is due to the fact that the time between two successive handoffs is assumed to be exponentially distributed in the CTMC model, which is not very accurate in real systems.

6.6.3 Availability Improvement

The unavailability for the three data communication systems when the train velocity changes from 20 to 100 km/h is shown in Figure 6.11. For comparison, we

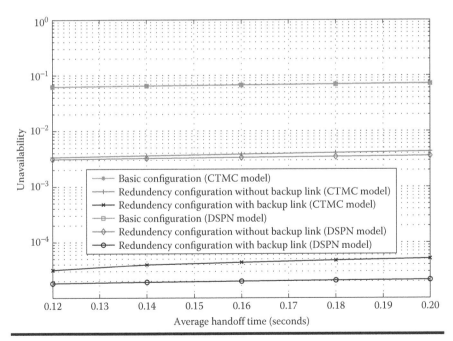

Figure 6.10 Comparison of CTMC and DSPN model solutions for different redundancy configurations.

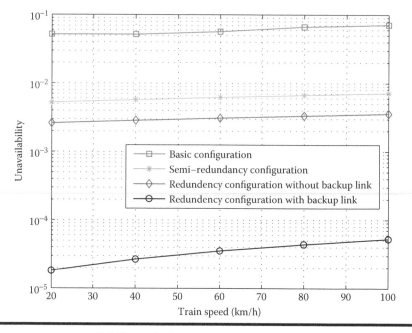

Figure 6.11 **Unavailability of the three WLAN-based data communication systems.**

calculate the unavailability of the system with semi-redundancy studied in [5], where only one SA is installed on the train. As shown in the figure, the system unavailability increases with the train velocity. This is because the larger the velocity, the more frequently the handoff occurs, which increases the handoff rate. Consequently, the unavailability of all the three systems is increased. Particularly, the unavailability of the existing system with basic configuration is more than 5% with different train speeds. Such a high unavailability would not be acceptable in practice. By contrast, the proposed data communication systems can decrease the unavailability below 1% with a wide range of the train velocity. From Figure 6.11, we can also observe that the second proposed data communication system with redundancy and backup link has better availability than the first one, although handoffs occur more frequently in the second system. This is because the probability of successful transitions from a backup link to an active link is much higher than the probability of deep channel fading. When a handoff happens at one of the active links, the corresponding backup link will become active before the other active link encounters deep channel fading. Therefore, the system will become unavailable only when all the four links fail. Our proposed redundancy configurations always give better availability performance compared to the semi-redundancy scheme as well.

6.7 Conclusion

Service availability is an important issue in CBTC systems. In this chapter, we proposed two WLAN-based data communication systems with redundancy configurations to improve the CBTC communication availability. We modeled each system as a CTMC. In order to check the soundness of the CTMC model, the data communication system behavior was modeled with DSPN. The availability of different data communication systems was compared. The results also show that the two proposed redundancy configurations significantly improve system availability, and the redundancy configuration with the backup link at the head and tail on the train achieves the best performance.

References

1. M. Aquado, E. Jacob, P. Saiz, J. J. Unzilla, M. V. Hiquero, and J. Matias. Railway signaling systems and new trends in wireless data communication. In *Proceedings of the IEEE VTC'2005-Fall*, Dallas, TX, September 2005.
2. F. Whitwam. Integration of wireless network technology with signaling in the rail transit industry. *Alcatel Telecommunications Review*, 1(1):43–48, 2003.
3. L. Zhu, F. R. Yu, B. Ning, and T. Tang. Cross-layer handoff design in MIMO-enabled WLANs for communication-based train control (CBTC) systems. *IEEE Journal on Selected Areas in Communication*, 30(4):719–728, 2012.
4. P. Bellavista, M. Cinque, D. Cotroneo, and L. Foschini. Self-adaptive handoff management for mobile streaming continuity. *IEEE Transactions on Network and Service Management*, 6(2):80–94, 2009.
5. T. Xu, T. Tang, C. Gao, and B. Cai. Dependability analysis of the data communication system in train control system. *Science in China*, 52-9:2605–2618, 2009.
6. E. Kuun. Open standards for CBTC and CCTV radio based communication. *Technical Forums of Alcatel*, 2(1):99–108, 2004.
7. R. Lardennois. Wireless communication for signaling in mass transit. *Siemens Transportation Systems*, 2003.
8. A. Leon-Garcia. *Probability and Random Processes for Electrical Engineering*. 2nd edition, Prentice Hall, NJ, 1994.
9. M. Ghaderi and R. Boutaba. Call admission control in mobile cellular networks: A comprehensive survey. *Wireless Communications and Mobile Computing*, 6(1):69–93, 2006.
10. J. Liu, Y. Yi, A. Proutiere, M. Chiang, and H. V. Poor. Towards utility-optimal random access without message passing. *Wireless Communications and Mobile Computing*, 10(1):115–128, 2010.
11. R. Zurawski and M. C. Zhou. Petri nets and industrial applications: A tutorial. *IEEE Trans. Industrial Electronics*, 41(6):567–583, 1994.
12. G. Egeland and P. Engelstad. The availability and reliability of wireless multi-hop networks with stochastic link failures. *IEEE Journal on Selected Areas in Communication*, 27(7):1132–1146, 2009.
13. R. Fantacci, D. Marabissi, and D. Tarchi. A novel communication infrastructure for emergency management: the In. Sy. Eme. vision. *Wireless Communications and Mobile Computing*, 10(12):1672–1681, 2010.

14. Y. Ma, J. J. Han, and K. S. Trivedi. Composite performance and availability analysis of wireless communication networks. *IEEE Transactions on Vehicular Technology*, 50(5):1216–1223, 2001.
15. W. Hneiti and N. Ajlouni. Dependability analysis of wireless local area networks. In *Proceedings of the Information and Communication Technologies*, Damascus, Syria, pages 2416–2422, April 2006.
16. D. Y. Chen, G. Sachin, and K. Chandra. Dependability enhancement for IEEE 802. 11 wirelesses LAN with redundancy techniques. In *Proceedings of the International Conference on Dependability Systems and Networks*, San Francisco, CA, pages 521–528, June 2003.
17. A. Zimmermann and Hommel G. Towards modeling and evaluation of ETCS real-time communication and operation. *Journal of Systems and Software*, 77:47–54, 2005.
18. H. Holger and J. David. A comparative reliability analysis of ETCS train radio communications. AVACS Technical Report, 2005.
19. J.-C. Lnprie. Dependable computing and fault tolerance: Concepts and terminology. In *Proceedings of the 15th International Symposium on Fault-Tolerant Computing*, Ann Arbor, MI, pages 2–7, June 1985.
20. A. Kleyner and V. Volovoi. Application of Petri nets to reliability prediction of occupant safety systems with partial detection and repair. *Reliability Engineering and System Safety*, 95(6):606–613, 2010.
21. D. C. Ionescu, A. P. Ulmeanu, A. C. Constantinescu, and I. Rotaru. Reliability modelling of medium voltage distribution systems of nuclear power plants using generalized stochastic Petri nets. *Computers and Mathematics with Applications*, 51(2):285–290, 2006.
22. J. L. Peterson. *Petri Net Theory and the Modeling of Systems*. Englewood Cliffs, NJ: Prentice Hall, 1981.
23. M. A. Marson and G. Chiola. On Petri nets with deterministic and exponentially distributed firing times. *Advances in Petri Nets*, 266:132–145, 1987.
24. H. Choi, V. Kulkarni, and K. Trivedi. Transient analysis of deterministic and stochastic Petri nets. In *Application and Theory of Petri Nets*, Chicago, IL, pages 166–185. 1993.
25. V. G. Kulkarni. *Advances in Petri Nets*. Lecture Notes on Stochastic Models in Operations Research. University of North Carolina, Chapel Hill, NC, 1990.

Chapter 7

Novel Communications-Based Train Control System with Coordinated Multipoint Transmission and Reception

Li Zhu, F. Richard Yu, and Tao Tang

Contents

7.1 Introduction 116
7.2 CBTC Systems 119
    7.2.1 Impacts of Communication Latency on CBTC Systems 119
    7.2.2 Proposed CBTC System with CoMP 121
7.3 System Models 122
    7.3.1 Train Control Model 123
    7.3.2 Communication Channel Model 125
7.4 Communication Latency in CBTC Systems with CoMP 126
    7.4.1 Coordinated Multipoint Transmission and Reception 126
    7.4.2 Data Transmission Rate and BER 126
    7.4.3 Communication Latency 128

7.5   Control Performance Optimization in CBTC Systems with CoMP.........129
  7.5.1   SMDP-Based CoMP Cluster Selection and Handoff
          Decision Model 129
          7.5.1.1   Decision Epochs 130
          7.5.1.2   Actions 130
          7.5.1.3   States 131
          7.5.1.4   Reward Function 131
          7.5.1.5   State Transition Probability 132
          7.5.1.6   Constraints..134
  7.5.2   Solutions to SMDP-Based CoMP Cluster Selection and
          Handoff Decision Scheme 134
          7.5.2.1   Reduced-State Bellman Equation 135
          7.5.2.2   Online Value Iteration Algorithm via Stochastic
                    Approximation...135
  7.5.3   Optimal Guidance Trajectory Calculation 137
7.6   Simulation Results and Discussions 139
  7.6.1   Train Control Performance Improvement.....................................140
  7.6.2   Handoff Performance Improvement 143
7.7   Conclusion 145
References 146

7.1 Introduction

In existing train–ground communication systems, the mobile terminal (MT) on a train communicates with an independent wayside base station to realize train–ground communications. Trains travel fast on the railways, and the received signal-to-noise ratio (SNR) changes rapidly. The communication latency will be a serious problem when the MT on a train is in deep fading. More importantly, when a train moves away from the coverage of a base station and enters the coverage of another base station along the railway, a handoff procedure occurs. In the current communications-based train control (CBTC) systems, only the traditional hard handoff scheme is supported, where a train can only communicate with a single base station at any given time. It may result in communication interruption and long latency due to the weak wireless signals in the handoff zone and high moving speed of trains. Both of these two challenges can severely affect train control performance, train operation efficiency, and the utilization of railway.

Recently, some research has been done on the train–ground communication issues in the railway environment. In [1], a fast handover algorithm suitable for dedicated passenger line is proposed by setting a new neighboring list. A novel handover scheme based on on-vehicle antennas is introduced in [2]. A cross-layer handoff design is studied in [3] for multiple-input and multiple-output (MIMO)-enabled wireless local area networks (WLANs).

Although these above works consider the impacts of railway environment on the communication performance, the handoff schemes in most current research are still hard handoff based, where the "break-before-make" principle limits the handoff performance improvement. In addition, the impacts of handoff latency on the control performance of CBTC systems is largely ignored in the existing works.

In this chapter, we use recent advances in coordinated multipoint (CoMP) transmission/reception to enable soft handoff, and consequently enhance the performance of CBTC systems. CoMP is a new method that helps with the implementation of dynamic base station coordination in practice. It is considered as a key technology for future mobile networks and is expected to be deployed in the future long-term evolution-advanced (LTE-A) systems to improve the cellular network performance [4]. To the best of our knowledge, using CoMP in CBTC systems has not been studied in previous works.

Intuitively, CoMP can improve not only the performance of commercial networks (e.g., cellular networks) also the performance of CBTC systems. Nevertheless, the adoption of CoMP in CBTC is not trivial due to the following reason: Traditional design criteria, such as network capacity, are used in existing works in CoMP-based networks. However, recent studies in cross-layer design show that maximizing capacity does not necessarily benefit the application layer [5–8], which is train control in CBTC systems. From a CBTC perspective, the performance of train control is more important than that at other layers. A commonly used control performance measure is the linear quadratic cost [9], which is directly related to train control accuracy, train safety, and passengers ride quality [10].

In this chapter, we propose a CBTC system with CoMP to enhance the train control performance of this system. The distinct features of this chapter are as follows:

1. We propose a CoMP-enabled CBTC train–ground communication system. With CoMP, a train can communicate with a cluster of base stations simultaneously, which is different from the current CBTC systems, where a train can only communicate with a single base station at any given time.

2. Unlike the existing works on communication systems that use capacity as the performance measure, in this chapter, linear quadratic cost for the train control performance in CBTC systems is considered as the performance measure.

3. We jointly consider CoMP cluster selection and handoff decision issues in CBTC systems. Although some works have been done to address the handoff problem, most of them focus on handoff protocols, and consequently handoff decision policy issues (i.e., when to perform handoff) are largely ignored in CBTC systems, which should be carefully considered. The handoff decision problem becomes more complicated when CoMP is used in CBTC, because the system needs to decide not only when to perform handoff but also which cluster to use.

4. In order to mitigate the impacts of communication latency on the train control performance, we propose an optimal guidance trajectory calculation scheme in the train control procedure that takes full consideration of the tracking error caused by communication latency.

5. The system optimization of CBTC system with CoMP is formulated as a semi-Markov decision process (SMDP) [11], which has been successfully used to solve opportunistic spectrum access in cognitive networks [12], among others. This chapter focuses on the application of SMDP to the control performance improvement in CBTC systems with CoMP.

6. Extensive simulation results are presented. It is shown that train control performance can be improved substantially in our proposed CBTC system with CoMP.

The main notations used in this chapter are summarized in Table 7.1.

Table 7.1   Main Notations

Notation	Definition	Unit
T	Communication period	s
$q(k)$	Train position	m
$v(k)$	Train velocity	m/s
$\varepsilon(k)$	Velocity tracking error	m/s
M	Train Mass	kg
$w_i(k)$	Slope resistance	N
$w_r(k)$	Curve resistance	N
$w_w(k)$	Wind resistance	N
$u(k)$	Train control command	N
$x(k)$	Train states	m, m/s
$y(k)$	Controller input	m, m/s
$x_c(k)$	Controller state	m/s
$z(k)$	Observed train states	m/s
α	Deceleration of the ATP braking curve	m/s²
$C(\theta)$	Channel capacity under cluster θ	bits/s
P_o	Transmitted power of each user terminal	mW
PR_{Td}	Channel transition probability after latency T_d	
$h_{B*}(k)$	Channel gain from a base station to the MT	dB
T_l	Current communication latency	ms
T_{afl}	Communication latency from the front train to the ground	ms

The rest of this chapter is organized as follows: Section 7.2 introduces CBTC systems. Section 7.3 describes the system models. Section 7.4 discusses the derivation of communication latency. Section 7.5 describes control performance optimization. Section 7.6 presents simulation results and discussions. Finally, Section 7.7 concludes this chapter.

7.2 CBTC Systems

In this section, we first present the impacts of wireless communications on CBTC systems that will be introduced next. Then, we describe the proposed CBTC system with CoMP.

7.2.1 Impacts of Communication Latency on CBTC Systems

The impacts of wireless communications on CBTC can be categorized into two aspects: safety and efficiency. We first give a simple example about the safety. Based on the locations of all the trains and other obstacles along the railway, the zone controller (ZC) transmits a movement authority (MA) to each train. One kind of obstacles is the switch. When a train approaches to a switch area, the ZC that controls this area will send the switch state to the train. When the switch state is in a normal state, the train will go through the switch to the normal direction without decreasing speed. However, the switch state may change to reverse or unknown due to equipment failure or human error. If the wireless link is not available due to communication latency at this moment, the ZC will not be able to report the updated switch state to the train. With long communication latency, the train will then go through the switch to the reversed direction with full speed, and an accident may happen.

Next we present the impacts on CBTC efficiency. In every communication period, the train and ground equipment sends the corresponding information to each other after processing the received information. As shown in Figure 7.1a, when there is no communication latency, Train 2 gets the updated MA in every communication period. The MA moves forward with Train 1. The velocity of Train 2 will not reach the target velocity on the braking curve when the two trains travel along the same guidance trajectory. Due to communication latency in train–ground communication systems, Train 2 may not be able to get the updated MA from ZC for a certain period of time. As shown in Figure 7.1b, Train 2 will still use the old MA before the updated MA arrives, and its automatic train protection (ATP) subsystem will start service brake to protect the train from traveling out of its MA. When the ATP subsystem receives the updated MA, it stops the service brake. The automatic train protection (ATO) system will then take control of the train and bring the train back to the original optimized guidance trajectory.

To make it easier to understand the impact of communication latency on CBTC efficiency, we have added a simulation result. We only simulate the communication

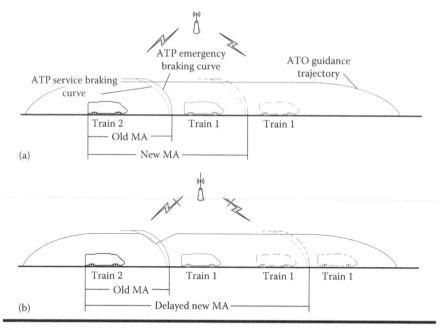

Figure 7.1 Impacts of wireless communications on CBTC efficiency. (a) Without communication latency. (b) With communication latency.

latency impact on CBTC efficiency when handoff happens. The communication latency impact when handoff does not happen is ignored. This is because the communication latency under this circumstance is only several milliseconds, and it is less than the communication period. In the simulation scenario, there are three trains traveling between two stations. The wayside base stations are deployed along the railway line with an average distance of 600 m between two successive base stations. The three trains depart from the station A successively with 16 s interval and stop at the station B. The communication period is 200 ms, the start acceleration is 1 m/s², the deceleration that defines the ATP service braking curve is set to be 1.2 m/s², the rail limited speed is 80 km/s, the train length is 100 m, and the system response time is 100 ms.

Based on the train and rail parameters, we first calculate the travel time between stations when the communication latency impact is not considered. The distances between two stations are set to be 2000, 3000, 4000, 5000, and 6000 m, respectively. Figure 7.2 shows the trip error between the third departed train and the first departed train when the handoff communication latency is 2, 4, and 6 s. As shown in the figure, the communication latency caused by handoff would lead to second trip time error. The ATS subsystem needs to adjust the timetable to make up the wasted time in the last station sections, which severely decreases the utilization of whole railway network infrastructure. In order to mitigate these impacts, we

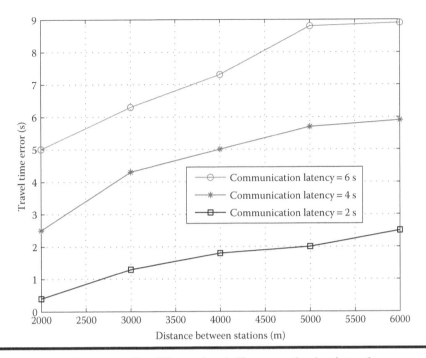

Figure 7.2   Trip error under different handoff communication latencies.

propose a CBTC train–ground communication system with CoMP, which will be presented in Section 7.2.2.

7.2.2 Proposed CBTC System with CoMP

The proposed CBTC system with CoMP is shown in Figure 7.3. Unlike the existing CBTC system, in the proposed system, a train can communicate with a cluster of base stations (BS1–BS4) (e.g., clusters C1–C5 in Figure 7.3) simultaneously, which is different from the current CBTC systems, where a train can only communicate with a single base station at any given time. When a base station fails, the remaining base stations in the cluster can still guarantee the availability of train–ground communications.

In the above system, as trains travel on the railway, the received SNR changes rapidly. The communication latency will be a serious problem when the MT on the train is in deep fading. In theory, CoMP communication can extend cell coverage and increase capacity in wireless networks. However, because the channel state information from the involved base stations is needed in CoMP systems, practical systems that employ CoMP techniques suffer from constraints imposed by the backhaul network, which is used to exchange the channel state information from the involved base stations. Backhaul networks are constrained

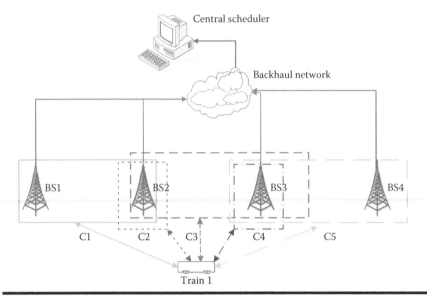

Figure 7.3 Proposed CBTC system with CoMP.

in capacity and introduce lost and/or outdated channel state information, which will result in performance degradation of CoMP systems. Therefore, we need to decide whether or not to use CoMP considering the potential quality-of-service (QoS) gain of CoMP and QoS degradation caused by the backhaul infrastructure latency.

More importantly, compared with CoMP-based commercial cellular systems, handoffs happen quite frequently in CBTC systems. Therefore, one of the critical issues in the system is the handoff decision (i.e., when to perform handoff) problem. If the handoff decision policy is not designed carefully, ping-pong effect and long communication latency may occur, which will significantly affect the performance of a CBTC system. Therefore, an efficient handoff decision policy is needed to decide at what time to trigger a handoff and whether or not CoMP communication should be used. In addition, due to the handoff latency in the system, it is also desirable to mitigate the impacts of handoff latency on the system control performance.

7.3 System Models

In this section, we describe the system models that will be used in our train control performance optimization. We first present the train control model. The communication channel model is described next.

7.3.1 Train Control Model

Figure 7.4 presents the train control model used in the chapter. The ATP subsystem first receives the MA (that has the front train states) sent from the ZC through the wireless link. It then calculates the braking curve for the received MA. The braking command will be executed when the train velocity is greater than the target velocity on the braking curve. Otherwise, the train is controlled by the ATO system. The ATO subsystem gets train states (velocity, position, etc.) from the ATP subsystem. It gives traction or braking command to bring the train to the guidance trajectory after comparing the train travel trajectory with the optimized guidance trajectory.

According to the train dynamics, the train state space equation can be written as [3]

$$
\begin{cases}
q(k+1) = q(k) + v(k) \cdot T + \dfrac{1}{2}\dfrac{u(k)}{M}T^2 \\
\qquad\qquad - \dfrac{1}{2}\dfrac{w_i(k) + w_r(k) + w_w(k)}{M}T^2 \\
v(k+1) = v(k) + \dfrac{u(k)}{M}\cdot T - \dfrac{w_i(k) + w_r(k) + w_w(k)}{M}T
\end{cases}
\tag{7.1}
$$

where:
T is the sampling rate, which depends on the communication period
$q(k)$ is the train position at time k
$v(k)$ is the train velocity
M is the train mass
$w_i(k)$ is the slope resistance

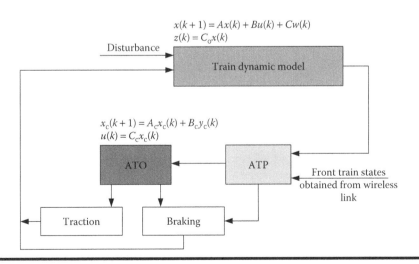

Figure 7.4   Train control model.

$w_r(k)$ is the curve resistance
$w_w(k)$ is the wind resistance
$u(k)$ is the train control command from the train controller

CBTC systems use packet-based transmission of train control information. Compared to traditional track-based train control systems, where the train only reports its location to the ground equipment and receives the information from ground equipment in specific places, in CBTC systems, the train sends its location and receives the MA from ground ZC continuously. In real system implementations, both information processing and transmission need time. Therefore, data packets are exchanged between the ground ZC and trains on a poll-response basis with a typical cycle time, which is the communication period in our chapter.

Equation 7.1 can be rewritten as

$$x(k+1) = Ax(k) + Bu(k) + Cw(k) \tag{7.2}$$

where:
$x(k) = \{q(k)\ v(k)\}$ is the state space
$w(k) = w_i(k) + w_r(k) + w_w(k)$ is the extra resistance acting on the train

The train dynamics model described in Equation 7.1 has been widely used in optimal train control studies [13,14]. It is shown that the optimal control problem for a train with a distributed mass on a track with a continuously varying gradient can be replaced by an equivalent problem for a point mass train, and that any strategy of continuous control can be approximated as closely as a strategy with discrete control.

We assume that the train controller is linear time invariant in discrete time and has the following state space model [3]:

$$\begin{cases} x_c(k+1) = A_c x_c(k) + B_c y(k) \\ u(k) = C_c x_c(k) \end{cases} \tag{7.3}$$

where:
$y(k)$ is the controller input, which includes the states of the two trains
$x_c(k)$ is the state of the controller

In this chapter, we use velocity tracking error as the state of the controller. Linear controllers have been successfully used in train control systems. For example, in [15], a linear controller is used in high-speed trains, where the commander of the controller is linearly and invariantly proportional to system states. Moreover, in this chapter, the SMDP-based optimization algorithm is not dependent on a specific train control model. In other words, other more sophisticated train control models can be used in our algorithm as well.

We consider the linear quadratic cost as our performance measure in this chapter. Specifically, the controller described in Equation 7.3 is designed to minimize

$$J_{LQ} = \sum_{k \to \infty} \left\{ \left[z(k) - \tilde{z}(k) \right]' Q \left[z(k) - \tilde{z}(k) \right] + u'(k) R u(k) \right\} \qquad (7.4)$$

where:

$z(k) = C_o x(k)$ is the observed train states

$\tilde{z}(k)$ is the destination train states

The weight matrix Q is positive semidefinite

R is positive definite

We can tune the system performance by choosing different Q and R. The first part in Equation 7.4 indicates the minimization of state tracking error, and the second part considers the control magnitude. Note that the square root of this performance measure in Equation 7.4 is equivalent to the H_2 norm for systems with perfect feedback [9]. We will refer to the performance measure as the H_2 norm due to this equivalence.

The optimal control can be realized only if the MA is updated when the train moves forward. The service brake started by the ATP subsystem due to communication latency will make the train travel off the optimized guidance trajectory, which degrades the system control performance.

7.3.2 Communication Channel Model

In this chapter, we use finite-sate Markov channel (FSMC) models in CBTC systems. FSMC models have been widely accepted in the literature as an effective approach to characterize the correlation structure of the fading process, including satellite channels [16], indoor channels [17], Rayleigh fading channels [18,19], Rician fading channels [20], and Nakagami fading channels [21,22]. Considering FSMC models may enable substantial performance improvement over the schemes with memoryless channel models [23].

In FSMC models, the range of the received SNR can be partitioned into discrete levels. Each level corresponds to a state in the Markov chain. Assume that there are L states in the model. Let i and γ denote the instantaneous channel state and SNR, respectively. When the channel is in state i, the corresponding SNR is γ_i. Then we have $\gamma_i < \gamma < \gamma_{i+1}$, $1 \leq i \leq L-1$. The probability of transition from state i to state j in the Markov model is denoted by $P_{i,j}$. In real systems, the values of the above transition probability can be obtained from the history observation of the CBTC system.

7.4 Communication Latency in CBTC Systems with CoMP

In this section, we derive the communication latency in CBTC systems with CoMP. We first derive the outage capacity and bit error rate (BER). Then, we obtain the communication latency.

7.4.1 Coordinated Multipoint Transmission and Reception

In cellular networks, interference exists between intercells, which affects spectral efficiency, especially in urban cellular systems. CoMP, which was originally proposed to overcome this limitation, can significantly improve average spectral efficiency and also increase the cell edge and average data rates.

In our system, CoMP is used among the active base stations to improve train–ground communication performance. In the downlink, the neighboring active base stations transmit cooperatively, and thus their coverage increases. In addition, diversity gain can be realized when the MT on the train combines the received signals from multiple base stations. In the uplink, multiple active base stations receive in coordination, which effectively reduces the requirement of received signal power at each individual base station. Together, CoMP can provide coverage for train MT in nearby cells that are under deep fading and improve the system availability.

In order to achieve the optimal performance, all of the base stations in the network should cooperate with each other in each transmission or reception process. However, the introduced complexity is not acceptable in real systems. Therefore, standards have specified the maximal number of base stations that may cooperate with each other [24]. In this chapter, we assume that there are only two base stations in each CoMP cooperation cluster. This is because the base stations are linearly deployed in CBTC train–ground communication systems. The base stations in the cluster can be switched on in a combination set $\Theta = \{1,2\}$, and each combination θ is an element of set Θ.

7.4.2 Data Transmission Rate and BER

We first derive the system data transmission rate. For an arbitrary element $\theta \in \Theta$, the uplink sum capacity for the cluster $C(\theta)$ can be calculated as follows [25]:

$$C(\theta, H) = \log_2 \det(I_{|\theta|} + P_o HH^+) \tag{7.5}$$

where:
$I_{|\theta|}$ denotes a $|\theta| \times |\theta|$ identity matrix
P_o denotes the transmission power of each user terminal
$H \in C^{|\theta| \times |\theta|}$ denotes the channel matrix

Once a cluster has been selected, the data rate allocation can be performed. We select the modulation and coding scheme for the train MT. The corresponding data rate can be defined as

$$\text{Rate} = \beta \cdot C(\theta, H) \tag{7.6}$$

where:
  β can be chosen from the interval $[0,1]$

Equation 7.5 works only if the channel matrix can be obtained with perfect accuracy by the central scheduler, which is difficult to be realized in most wireless systems. This is because CoMP systems suffer from constraints that are imposed by the backhaul infrastructure, which is required to communicate with the central scheduler and exchange channel information for joint processing.

In addition to the train control information, passenger information system, public security monitor system, and station-to-train audio dispatch system are all integrated into a single backhaul network. In a backhaul network with so much traffic, the communication congestion and latency are unavoidable in the backhaul network. This is why the CoMP suffers from constraints that are imposed by the backhaul infrastructure of CBTC systems.

Given the Markov channel model and the current observed channel state H, the channel state will transition to $\hat{H}$ with probability $PR_{T_d}(H, \hat{H})$ after an approximately backhaul infrastructure latency T_d. $PR_{T_d}(H, \hat{H})$ is the channel transition probability, which can be obtained from field tests. The system capacity $C(\theta, \hat{H})$ under channel state $\hat{H}$ can be calculated as described in Equation 7.5.

If the system capacity $C(\theta, \hat{H})$ is less than the allocated data rate, there is an outrage and the resulted data rate is 0. Otherwise the resulted data rate Râte is equal to the allocated data rate Rate. Therefore, we have

$$\begin{cases} \text{Râte} = 0, \text{ if } C(\theta, \hat{H}) < \text{Rate} \\ \text{Râte} = \text{Rate, others} \end{cases} \tag{7.7}$$

The outrage probability P_{out} for the observed channel state H and allocated data rate Rate can be approximately derived as

$$P_{\text{out}}(H, \text{Rate}) = \sum_{\hat{H}} PR_{T_d}(H, \hat{H}), \text{ if } C(\theta, \hat{H}) < \text{Rate} \tag{7.8}$$

As a result, the ultimate data rate can be calculated as

$$\text{Râte} = \text{Rate} \cdot [1 - P_{\text{out}}(H, \text{Rate})] \tag{7.9}$$

Next, we start to derive the BER. According to [26], when CoMP is not used in our proposed system, that is, the selected cluster is a single base station, communication over the Rayleigh channel from source to destination without diversity gain has the BER performance:

$$BER_1 = \frac{1}{2}\left[1 - \sqrt{\frac{\gamma}{1+\gamma}}\right] \qquad (7.10)$$

where:

γ is the received signal SNR, and it can be obtained from the transmission power and channel state

When neighboring active base stations transmit or receive cooperatively to obtain diversity gain, that is, the selected cluster has two base stations, we consider the maximum ratio combining (MRC) of two diversity branches to derive the system BER as [26]

$$BER_2(\gamma_1,\gamma_2) = \frac{1}{2}\left[1 + \frac{1}{\gamma_1 - \gamma_2}\left(\frac{\gamma_2}{\sqrt{1+(1/\gamma_2)}} - \frac{\gamma_1}{\sqrt{1+(1/\gamma_1)}}\right)\right] \qquad (7.11)$$

where:

γ_1 and γ_2 are the received signal SNRs for the two diversity branches, and they can be obtained from the transmission power and channel state

Similar to the circumstance regarding data transmission rate, when CoMP is used in the system, given the current observed channel states H_1 and H_2 for the two diversity branches, the channel state will transition to $\hat{H}_1$ and $\hat{H}_2$ with probability $PR_{T_d}(H_1,\hat{V}_1)$ and $PR_{T_d}(H_2,\hat{H}_2)$ after latency T_d, respectively. Therefore, given the current observed channel states H_1 and H_2, the real system BER can be calculated as

$$\hat{BER}_2 = \sum_{\hat{H}_1}\sum_{\hat{H}_2} PR_{T_d}\left(H_1\hat{H}_1\right) \times PR_{T_d}\left(H_1\hat{H}_1\right) \times BER_2\left(\gamma_1\hat{H}_1,\gamma_2\hat{H}_2\right) \qquad (7.12)$$

7.4.3 Communication Latency

In order to improve system reliability, an automatic repeat request (ARQ) scheme is needed. As selective-repeat (SR)-ARQ has been proven to outperform other basic forms of ARQ schemes (such as stop-and-wait ARQ and go-back-N ARQ) [27], we use SR-ARQ in this study. According to an SR-ARQ protocol, the average number of transmissions needed for one packet to be successfully accepted by the destination is

$$1 \cdot p_c + 2 \cdot p_c(1 - p_c) + 3 \cdot p_c(1 - p_c)^2 + \cdots = \frac{1}{p_c} \qquad (7.13)$$

where p_c is the probability that a packet is successfully received, and it can be calculated as

$$p_c = (1 - \mathrm{BER})^{L_{\mathrm{packet}}} \tag{7.14}$$

where:

BER is described in Equations 7.10 and 7.12
L_{packet} is the number of bits contained in one packet

With the average number of transmissions needed for one packet to be successfully accepted $1/p_c$, the communication latency can be approximately calculated as

$$T_{\mathrm{la}} = \frac{1}{p_c} \cdot \frac{L_{\mathrm{packet}} + L_{\mathrm{ack}}}{\hat{Rate}} \tag{7.15}$$

where:

L_{ack} is the ACK length
$\hat{Rate}$ is the data transmission rate

7.5 Control Performance Optimization in CBTC Systems with CoMP

As described in Section 7.2.1, communication latency in train–ground communication systems could make the train travel away from the guidance trajectory and severely affect CBTC system performance. In order to mitigate the impacts of communication latency on system performance, we jointly consider CoMP cluster selection and handoff decision issues in CBTC systems. Particularly, the optimal scheme is composed of two parts. For the first part, we propose an SMDP-based optimization algorithm, with the optimization objective to minimize the linear quadratic cost of train control system. In addition, when the train controller finds its speed deviates from a preset value due to communication latency, it will recalculate the optimal guidance trajectory again. The train will travel along the new guidance trajectory after the recalculation.

In this section, we first describe our SMDP-based model. The online value iteration algorithm via stochastic approximation to solve the SMDP model is introduced next. Finally, the optimal guidance trajectory calculation approach is described.

7.5.1 SMDP-Based CoMP Cluster Selection and Handoff Decision Model

With the objective to minimize the linear quadratic cost function, we model the train–ground communication optimization problem as an SMDP. We first present an overview of SMDP modeling. Then, the states, action, reward functions, state transition probability, and constraints are presented.

Markov decision process (MDP) provides a mathematical framework for modeling decision making in situations where outcomes are partly random and partly under the control of a decision maker. Besides the basic features, an SMDP generalizes an MDP by allowing decision maker to choose actions whenever the system state changes and allowing the time spent in a particular state to follow an arbitrary probability distribution.

In our proposed CBTC systems with CoMP, the MT on the train makes handoff and CoMP cluster selection decisions at specific time instances according to the current state $s(t)$, and the system moves into a new state based on the current state $s(t)$ as well as the chosen decision $a(t)$. Given $s(t)$ and $a(t)$, the next state is conditionally independent of all previous states and actions. This Markov property of state transition process makes it possible to model the CoMP cluster selection and handoff problem as an SMDP.

An SMDP model consists of the following six elements: (1) decision epochs, (2) actions, (3) states, (4) reward functions, (5) state transition probabilities, and (6) constraints, which will be described in Sections 7.5.1.1 through 7.5.1.6.

7.5.1.1 Decision Epochs

The MT on a train has to make a decision whenever a certain time period has elapsed. The instant times are called *decision epochs*, and we designate it as k.

7.5.1.2 Actions

In our SMDP model, at each decision epoch, the MT on the train first has to decide whether or not to perform handoff action. Afterward, the CoMP cluster should be determined. As shown in Figure 7.3, we assume the MT on the train will not be in the coverage of more than four successive base stations, and we denote them as $B1, B2, B3$, and $B4$. As a result, there are five potential clusters in the system. They are $(B1, B2), (B2), (B2, B3), (B3)$, and $(B3, B4)$, and we denote them as $C1, C2, C3, C4$, and $C5$, respectively.

In order to achieve the optimal performance, all of the base stations in the network should cooperate with each other in each transmission or reception process. However, the introduced complexity may not be acceptable in real systems. Therefore, standards have specified the maximal number of base stations that may cooperate with each other [24]. In CBTC systems, the base stations are linearly deployed. For ease of presentation, we choose only two base stations in the cluster, because a third base station is too far away from the mobile station, and the performance gain brought by adding a base station into the cluster may not be worth the complexity cost in real systems. We also need to point out that our model is not limited to a cluster of only one or two base stations. It can be easily extended to incorporate more than two base stations in a cluster.

These five clusters correspond to five different actions $a(k)$ that can be performed at each decision epoch. Therefore, the action space can be described as

$$A = \{C1, C2, C3, C4, C5\} \tag{7.16}$$

7.5.1.3 States

The composite state $s(k) \in S$ is given as

$$s(k) = \{h_{B1}(k), h_{B2}(k), h_{B3}(k), h_{B4}(k), \xi(k), \varepsilon(k)\} \tag{7.17}$$

where:

$h_{B1}(k)$, $h_{B2}(k)$, $h_{B3}(k)$, and $h_{B4}(k)$ are the channel gains from four successive base stations to the train MT, respectively

$\xi(k)$ is the currently used cluster

$\varepsilon(k)$ is the velocity tracking error which is the error between the current train velocity and the reference velocity on the guidance trajectory

$h_{B1}(k)$, $h_{B2}(k)$, $h_{B3}(k)$, and $h_{B4}(k) \in \{\gamma_1, \gamma_2 \cdots \gamma_L\}$, where L is the number of states in the channel model

$\xi(k) \in \{C1, C2, C3, C4, C5\}$, because the currently used cluster is completely determined by the current action

The velocity tracking error ε is obtained by comparing the current train velocity with the destination velocity, and the destination velocity is calculated based on the distance between the two trains. In order to make the SMDP solvable, $\varepsilon(k)$ is discretized, and $\varepsilon(k) \in \{\varepsilon_1, \varepsilon_2, \ldots, \varepsilon_K\}$, where K is the total number of velocity tracking error states.

7.5.1.4 Reward Function

Reward function reflects the reward that can be obtained under a certain state and action. With the objective to minimize linear quadratic cost function as shown in Equation 7.4, we define the reward function as the reciprocal of the sum of tracking error and control magnitude. The reward function is closely related to the communication latency in the CBTC system. We need to point out that our proposed CBTC system with CoMP is not limited to a certain type of wireless technology. Once the communication latency is obtained, it can also be used in CBTC systems with dedicated short-range communications [28], LTE, or WiMax.

We present the rewards under different circumstances based on the communication latency model described in Section 7.4 as follows:

When $a(k) = \xi_k$, which means that the currently used cluster does not change and no handoff happens, the reward under this circumstance is defined as

$$r\left[h_{B1}(k), h_{B2}(k), h_{B3}(k), h_{B4}(k), \xi(k), \varepsilon(k), a(k)\right] =$$

$$\begin{cases} 1/\left\{Q\left[\varepsilon(k) + \alpha \times T + \alpha \times T_{afl}\right]^2 + Ru^2\right\}, & \text{if } T < T_1 \\ \\ 1/\left\{Q\left[\varepsilon(k) + \dfrac{u}{M} \times T + \alpha \times T_{afl}\right]^2 + Ru^2\right\}, & \text{if } T > T_1 \end{cases}$$

(7.18)

where:

u is the control command from the ATO subsystem to bring the train to the optimized guidance trajectory

T is the communication period

M is the train mass

α is the deceleration that defines the ATP service braking curve

T_1 is the current train–ground communication latency, which can be calculated with current channel state $h_{B1}(k), h_{B2}(k), h_{B3}(k), h_{B4}(k)$ according to the communication latency model

When $T < T_1$, the current communication latency is greater than the communication period and the MA cannot be updated under this decision epoch. Therefore, the velocity tracking error is increased by $\alpha \cdot T$ due to a communication interruption of T. For any train that is traveling behind, when the front train is in long latency/interruption state, it cannot receive the updated MA. This is because the train in front cannot report its new position to ZC. Given the average communication latency from the front train to ground T_{afl}, the velocity tracking error increased by the front train latency is $\alpha \cdot T_{afl}$.

When $a(k) \neq \xi_k$, which means that a base station joins or leaves the currently used cluster and a handoff happens, the communication latency between the train and the ground will be increased by $T_1 + T_{ex}$. T_{ex} is due to the extra signal exchange. The reward under this circumstance is defined as

$$r\left[h_{B1}(k), h_{B2}(k), h_{B3}(k), h_{B4}(k), \xi(k), \varepsilon(k), a(k)\right] =$$

$$1/\left\{Q\left[\varepsilon(k) + \alpha(T_{ex} + T_1) + \alpha \times T_{afl}\right]^2 + Ru^2\right\}$$

(7.19)

7.5.1.5 State Transition Probability

Given the current state, $s(k) = \{h_{B1}(k), h_{B2}(k), h_{B3}(k), h_{B4}(k), \xi(k), \varepsilon(k)\}$, and the chosen action, $a(k)$, the probability function of the next state, $s(k+1) = \{h_{B1}(k+1), h_{B2}(k+1), h_{B3}(k+1), h_{B4}(k+1), \xi(k+1), \varepsilon(k+1)\}$, is given by $P\left[s(k+1) \mid s(k), a(k)\right]$. Here, for simplicity of formulation and presentation, we assume that the wireless channels, currently used cluster, and velocity tracking error are independent. This assumption is reasonable in practice, because the wireless channels for different links are independent, and the currently used cluster is solely determined by the last action. Moreover, the velocity tracking error is dependent on the train dynamic model, which makes it reasonable to assume that it is independent from other components in the state. Therefore, we have

$$P(s(k+1)\,|\,s(k),a(k)) = P[h_{B1}(k+1)\,|\,h_{B1}(k)] \cdot P[h_{B2}(k+1)\,|\,h_{B2}(k)] \cdot$$

$$P[h_{B3}(k+1)\,|\,h_{B3}(k)] \cdot P[h_{B4}(k+1)\,|\,h_{B4}(k)] \cdot$$

$$P[\xi(k+1)\,|\,\xi(k),a(k)] \cdot P[H(k+1)\,|\,H(k),a(k)] \cdot \quad (7.20)$$

$$P[\varepsilon(k+1)\,|\,\varepsilon(k),a(k)]$$

where:

$P[h_{B1}(k+1)\,|\,h_{B1}(k)]$, $P[h_{B2}(k+1)\,|\,h_{B2}(k)]$, $P[h_{B3}(k+1)\,|\,h_{B3}(k)]$, and $P[h_{B4}(k+1) \,|\, h_{B4}(k)]$ are the channel state transition probabilities for different wireless links, respectively

$P[\xi(k+1)\,|\,\xi(k),a(k)]$ is the currently used cluster transition probability

$P[\varepsilon(k+1)\,|\,\varepsilon(k),a(k)]$ is the velocity tracking error transition probability

The channel state transition probabilities can be obtained from real field test data, which is described in Section 7.3.2. Other state transition probabilities will be derived in the following.

First, we derive the transition probability for the currently used CoMP cluster. Because the next used cluster is determined by the chosen action, the currently used cluster transition probability can be simply derived as

$$P[\xi(k+1)\,|\,\xi(k),\ a(k)] = \begin{cases} 0, & \text{if } a(k) \neq \xi(k+1) \\ 1, & \text{if } a(k) = \xi(k+1) \end{cases} \quad (7.21)$$

Second, we derive the transition probability for the velocity tracking error. The velocity tracking error is dependent on the control command at every decision epoch and the handoff action. Given a control command from the ATO subsystem to bring the train to the optimized guidance trajectory, the velocity tracking error transition probability is derived as

$$P[\varepsilon(k+1)\,|\,\varepsilon(k),\ a(k)] =$$

$$\begin{cases} 1, & \text{if } a(k) = s_k, T > T_1, \\ & \text{and } \varepsilon(k+1) = \varepsilon(k) + \alpha \times T + \alpha \times T_{\text{afl}} \\ 1, & \text{if } a(k) = s_k, T < T_1, \\ & \text{and } \varepsilon(k+1) = \varepsilon(k) + \dfrac{u}{M} \times T + \alpha \times T_{\text{afl}} \\ 1, & \text{if } a(k) \neq s_k, \\ & \text{and } \varepsilon(k+1) = \varepsilon(k) + \alpha(T_{\text{ex}} + T_1) + \alpha \times T_{\text{afl}} \\ 0, & \text{otherwise} \end{cases} \quad (7.22)$$

where:

u is the control command

T is the communication period

M is the train mass

T_l is the current train–ground communication latency

T_{afl} is the average communication latency from the front train to the ground

α is the deceleration that defines the ATP service braking curve

7.5.1.6 Constraints

As we described in Section 7.5.1.2, the handoff action in the system is performed by switching between different clusters. In order to avoid the "break-before-make" situation and realize soft handoff, action $a(k)$ is constrained in certain states. It can only switch between neighbor clusters. Mathematically, the constraints are defined as

$$a(k) \neq C_j, \quad j < i-1 \quad \text{if} \quad \xi(k) \neq C_i$$

and (7.23)

$$j > i+1 \quad \text{if} \quad \xi(k) = C_i$$

7.5.2 Solutions to SMDP-Based CoMP Cluster Selection and Handoff Decision Scheme

A decision rule prescribes a procedure for action selection in each state at a specified decision epoch. Markov decision rules are functions $\delta(k): S \rightarrow A$, which specify the action choice $a(k)$ when the system occupies state $s(k)$ at decision epoch k. A policy $\pi = (\delta(1), \delta(2), ..., \delta(k))$ is a sequence of decision rules to be used at all decision epochs.

A stationary control policy π induces a joint distribution for the random process $\{s(k)\}$. The optimal policy for SMDP can be obtained by solving the Bellman equation [11] recursively as

$$\rho + V(s) = \min_{\delta(s)} \left\{ r[s, \delta(s)] + \sum_{s'} Pr[s' \mid s, \delta(s)] V(s') \right\}$$ (7.24)

where:

$\delta(s)$ is the cluster selection action taken in state s

$r(s, \delta(s))$ given by Equations 7.18 and 7.19 is the per-stage reward when the current state is s and action $\delta(s)$ is taken

If there is a $(\rho, V(s))$ satisfying Equation 7.24, then ρ is the optimal average reward per stage, and the optimizing policy is given by $\pi^*(s) = \delta^*(s)$, where $\delta^*(s)$ is the optimizing action of Equation 7.24 at state s.

7.5.2.1 Reduced-State Bellman Equation

Instead of working on the global state space $s(k) = \{h_{B1}(k), h_{B2}(k), h_{B3}(k), h_{B4}(k), \xi(k), \varepsilon(k)\}$, we shall derive a reduced-state Bellman equation from Equation 7.24 using partitioning of the control policy π, which is based on partial system state ε only. Specifically, we partition a unichain policy π into a collection of actions as follows:

Definition (conditional actions): Given a policy π, we define $\pi(\varepsilon) = \{\pi(s): s = (h_{B1}, h_{B2}, h_{B3}, h_{B4}, \xi, \varepsilon) \ \forall h_{B1}, h_{B2}, h_{B3}, h_{B4}, \xi\}$ as the collection of actions under a given ε for all possible $h_{B1}, h_{B2}, h_{B3}, h_{B4}, \xi$. The policy π is therefore equal to the union of all the conditional actions, That is, $\pi = \bigcup_\varepsilon \pi(\varepsilon)$.

$$\rho + \tilde{V}(\varepsilon) = \min_{\delta(\varepsilon)} \left\{ \tilde{r}[\varepsilon, \delta(\varepsilon)] + \sum_{\varepsilon'} Pr[\varepsilon' \mid \varepsilon, \delta(\varepsilon)] V(\varepsilon') \right\} \quad (7.25)$$

where:

$\tilde{V}(\varepsilon) = E[V(s) \mid \varepsilon]$ is the conditional potential function

$\delta(\varepsilon) = \pi(\varepsilon)$ is the collection of actions under a given ε

$\tilde{r}[\varepsilon, \delta(\varepsilon)] = E\{r[s, \delta(s)]\}$ is the conditional per-stage reward

$\tilde{Pr}[\varepsilon' \mid \varepsilon, d(\varepsilon)] = E[Pr(s' \mid s, \delta(s) \mid \varepsilon)]$ is the conditional average transition kernel

A solution to this equation is still very complex due to the huge dimensionality of the state space, and brute force value iteration or policy iteration [11] has exponential memory size requirement. As a result, it is desirable to obtain an online and low-complexity solution to the problem.

7.5.2.2 Online Value Iteration Algorithm via Stochastic Approximation

In this section, we shall derive a low-complexity (but optimal) solution by proposing an online value iteration to solve the reduced-state Bellman equation in Equation 7.25. We shall propose an online sample path-based iterative learning algorithm to estimate the performance potential and the optimal policy. Define a vector mapping Π: with the ith component mapping $(1 \le i \le K)$ as

$$\Pi_i(\tilde{V}) = \min_{\delta(\varepsilon^i)} \left\{ \tilde{r}[\varepsilon^i, \delta(\varepsilon^i)] + \sum_{\varepsilon^j} Pr[\varepsilon^j \mid \varepsilon^i, \delta(\varepsilon^i)] V(\varepsilon^j) \right\} \quad (7.26)$$

where:

K is the total number of velocity tracking error states

Because the potential is unique up to a constant, we could set $\Pi_i(\tilde{V}) - \tilde{V}(\varepsilon^i) = J_\chi$ for a reference state ε^i $(1 \le i \le K)$. Let $\{\varepsilon(0), \ldots, \varepsilon(k), \ldots\}$ be the sample path, that is, the corresponding realizations of the system states. To perform the online value iteration, we divide the sample path into regenerative periods. A regenerative period is defined as the minimum interval that each ε state is visited at least once. Let $l_d(i)$ and $\tilde{V}_d$ be the times that ε^i is visited and the estimated performance potential in the dth regenerative period, respectively. Let $n_0 = 0$, $n_{d+1} = \min\{k+1 : k > n_d, \min_i l_d(i) = 1\}$ for $d \geq 0$. Then the sample path in the dth regenerative period is $\{\varepsilon(n_d), \ldots, \varepsilon(n_{d+1} - 1)\}$. At the beginning of the dth regenerative period ($n_d \leq t \leq n_{d+1} - 1$), initialize the following dummy variables as $S_{\tilde{g}_d}(i) = 0$, $S_{\tilde{V}_d}(i) = 0$, and $l_d(i) = 0$. Within the dth regenerative period, we adopt policy $\neq_d$. After observing the velocity error state $\varepsilon(k+1)$ at the end of the kth slot, update the following metric of the visited state $\varepsilon(k)$ according to

$$
\begin{cases}
S_{\tilde{g}_d}(i) = S_{\tilde{g}_d}(i) + r[s, \pi_d(s)] \\
S_{\tilde{V}_d}(i) = S_{\tilde{V}_d}(i) + \tilde{V}_d[\varepsilon(k)] & \text{if } \varepsilon(k) = \varepsilon^i \\
l_d(i) = l_d(i) + 1
\end{cases} \tag{7.27}
$$

At the end of the dth regenerative period, using stochastic approximation algorithm [29], we update the estimated potential for the $(d+1)$th regenerative period, which is

$$
\tilde{V}_{d+1}(\varepsilon^i) = \tilde{V}_d(\varepsilon^i) + \varsigma_d Y_d(\varepsilon^i) \tag{7.28}
$$

where:

$$
Y_d(\varepsilon^i) = \frac{S_{\tilde{g}_d}(i)}{l_d(i)} - \left[\frac{S_{\tilde{g}_d}(K)}{l_d(K)} + \frac{S_{\tilde{V}_d}(K)}{l_d(K)} - \tilde{V}_d(\varepsilon^K) \right] + \frac{S_{\tilde{g}_d}(i)}{l_d(i)} - \tilde{V}_d(\varepsilon^i) \tag{7.29}
$$

ς_d is the step size of the stochastic approximation algorithm
ε^K is the reference state

Without loss of generality, we set the state that the velocity track error with the highest value is the reference state. Accordingly, we update the policy for the $(d+1)$th regenerative period, which is given by

$$
\pi_{d+1} = \arg\min \Pi \, (\tilde{V}_{d+1}) \tag{7.30}
$$

Therefore, we could construct an online value iteration algorithm as follows:

Step 1 (initialization): Set $k = 0$ and start the system at an initial state $\varepsilon(0)$. Set $d = 0$ and initialize the potential $\tilde{V}_0$ and policy $\pi_0 = \arg\min T(\tilde{V}_0)$ in the 0th regenerative period.

Step 2 (online potential estimation): At the beginning of the dth regenerative period, set $S_{\tilde{g}_d}(i) = 0$, $S_{\tilde{V}_d}(i) = 0$, and $l_d(i) = 0$, $\forall i$. Run the system with policy π_d to $n_{d+1} - 1$ and accumulate the information of the visited ε from slot to slot according to Equation 7.27. At $n_{d+1} - 1$, update the estimated potential $\tilde{V}_{d+1}$ for the $(d+1)$th regenerative period according to Equation 7.28.

Step 3 (online policy improvement): Update the policy π_{d+1} for the $(d+1)$th regenerative period according to Equation 7.30.

Step 4 (termination): If $\|\tilde{V}_{d+1} - \tilde{V}_d\| < \sigma_v$, stop; otherwise, set $d := d+1$ and go to step 2.

7.5.3 Optimal Guidance Trajectory Calculation

An optimal CoMP cluster selection and handoff decision policy, which minimizes the linear quadratic cost function, can be derived from the algorithm presented above. Recall that the linear quadratic cost function is defined as the sum of tracking error and control magnitude in Equation 7.4. Due to the handoff latency in the system, the tracking error always exists when the train travels along the guidance trajectory. In order to further reduce the tracking error, the second part of our scheme is to recalculate the guidance trajectory. The newly calculated guidance trajectory should take full consideration of the tracking error caused by handoff latency. The calculation approach is presented in the following.

In traditional fixed-block track circuit-based train control systems, optimal guidance trajectory calculation has been studied for many years. The earliest and most noticeable work is from the Scheduling and Control Group in North Australia [30]. They conducted theoretic research and project on optimal train travel trajectory considering energy saving and trip time. The results show that the optimal guidance trajectory could be divided into four phases: traction, speed holding, coasting, and braking, as illustrated in Figure 7.5. The shift and velocity of four phases are S_1, S_2, S_3, S_4 and v_1, v_2, v_3, v_4, respectively. Based on that, many researchers investigated the optimal travel trajectory with the aim to find the switch point [30]. However, most existing works focus on traditional fixed-block track circuit-based train control systems, where the train travel trajectory is not affected by the front trains.

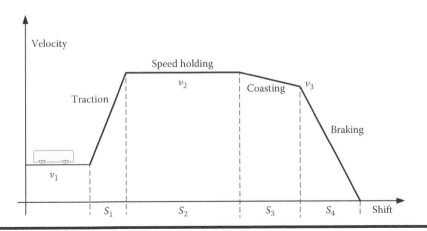

Figure 7.5 Optimal train guidance trajectory.

In this chapter, we propose an optimal guidance trajectory calculation in CBTC systems with CoMP. The scheme takes full consideration of the impacts from the front train. The scheme is described as follows:

Step 1: Initialize the basic parameters, train mass M, train traction power F_{trac}, train braking power F_{brac}, trip time T_{trip}, and trip distance S_{trip}. Set the current train velocity to be $v_1 = 0$, the current train position to be $S_{\text{cur}} = 0$, and the current traveled time to be $T_{\text{cur}} = 0$.

Step 2: Compute the remaining trip time, $T_{\text{remaintrip}} = T_{\text{trip}} - T_{\text{cur}}$, and the remaining trip distance, $S_{\text{remaintrip}} = S_{\text{trip}} - S_{\text{cur}}$. With $T_{\text{remaintrip}}$, v_1, $S_{\text{remaintrip}}$, and the basic parameters, the optimal guidance trajectory can be calculated as follows:

As shown in Figure 7.5, the optimal guidance trajectory could be divided into four phases: traction, velocity holding, coasting, and braking. Therefore, given the already known traction power and braking power, the optimal guidance trajectory can be determined if the holding velocity v_2 and the velocity v_3 that indicates the end of the costing phase are obtained. With this objective, the optimization problem can be formulated as

$$\min f = \frac{1}{2} M (v_2^2 - v_1^2) + F_{\text{fric}} \cdot S_2$$

s.t.

$$v_1 \le v_2 \le v_3$$

$$S_{\text{remaintrip}} = \frac{v_2^2 - v_1^2}{2 F_{\text{trac}}/M} + S_2 + \frac{v_3^2 - v_2^2}{2 F_{\text{fric}}/M} + \frac{v_3^2}{2 F_{\text{brak}}/M}$$

$$T_{\text{remaintrip}} = \frac{v_2 - v_1}{F_{\text{trac}}/M} + \frac{S_2}{v_2} + \frac{v_3 - v_2}{F_{\text{fric}}/M} + \frac{v_3}{F_{\text{brak}}/M}$$

(7.31)

In this equation, f is the energy consumed to accelerate the train from v_1 to v_2, and keep it travel at speed v_2 for a distance of S_2. The second constraint is the trip distance constraint, which includes four parts representing the distance traveled in the traction phase, the distance traveled in the velocity holding phase, the distance traveled in the coasting phase, and the distance traveled in the braking phase, respectively. The sum of these four distances should be equal to the remaining trip distance. The third constraint is the trip time constraint, which includes four parts representing the time traveled in the traction phase, the time traveled in the velocity holding phase, the time traveled in the coasting phase, and the time traveled in the braking phase, respectively. The sum of these four parts should be equal to the remaining trip time.

Step 3: If $S_{\text{remaintrip}} = 0$, go to step 4. Otherwise, compare the current train velocity v_1 with the reference train velocity v_{ref} on the calculated guidance trajectory. If $|v_1 - v_{\text{ref}}| \ge \Delta v$, where Δv is the preset velocity error, go to step 2.

Otherwise, update the train velocity v_1 and train position S_{cur} according to the train control model described in Section 7.3.1.

Step 4: The train stops at the destination. Wait until the train starts, and then go to Step 1.

7.6 Simulation Results and Discussions

In this section, simulation results are presented to illustrate the performance of our proposed system (Table 7.2). The system optimization process can be divided into the off-line part and the online part. For the off-line part, the mathematical models described earlier are used to derive the optimal policy for CoMP cluster selection and handoff decisions. The calculation of the optimal policy using the value iteration algorithm is performed off-line with C language. Once the optimal policy is obtained, it is stored in a table format. Each entry of the table specifies the optimal action (CoMP cluster selection and handoff) given the current state (i.e., channel state, currently used cluster).

For the online part, computer simulation based on NS2.29 is used. Specifically, at each decision epoch, each MT on the train looks up the policy table to find out the optimal action corresponding to its current state, and then executes action. Online looking up tables can be designed with little computational complexity in practice.

In the simulations using NS2, we consider a scenario with three moving nodes (MT on the trains) traveling between two stations, A and B. The wayside nodes (base stations) are deployed along the railway line with an average distance of 600 m between two successive base stations. These three moving nodes depart from station A successively with a constant interval and stop at station B. In the moving

Table 7.2 Simulation Parameters

Notation	Definition	Value
M	Train mass	50,000 kg
T_{ex}	Extra delay caused by handoff	9 µs
P_o	Transmission power	100 mW
L_{ack}	ACK packet size	20 bytes
L_{packet}	Data packet size	200 bytes
L	Number of channel states	4
K	Number of tracking error states	5
T	Communication period	100 ms

process, the moving nodes communicate with the wayside nodes. The nodes' moving speed is controlled by the train control model, and the communication latency is determined by the communication model in Section 7.4.3. Each node looks up the policy table to find out the optimal action corresponding to its current state, and then executes action. Static variables are collected in the simulations to obtain the train control performance H_2 norm, train travel trajectory, handoff policy, and the trip time.

We compare the performance of our proposed system with three other schemes. The first scheme is an existing scheme [31], where CoMP is not used. The train MT makes handoff decisions based on the immediate reward, and the optimal guidance trajectory is not recalculated when the speed deviation occurs. We denote this scheme as the *existing scheme*. For the second scheme, CoMP is used in the system, but the SMDP model is not used. Instead, the decision maker makes handoff decisions by the reward derived from the current channel state. The optimal guidance trajectory is not recalculated in this scheme. We denote this scheme as the *CoMP-based greedy scheme w/o updated trajectory*. For the third scheme, CoMP is used in the system, and the SMDP model is used to calculate the handoff decision policy. Unlike the proposed scheme, the optimal guidance trajectory is not recalculated in this scheme. We denote this scheme as the *CoMP-based scheme w/o updated trajectory*, and our proposed scheme as the *proposed scheme*.

7.6.1 Train Control Performance Improvement

We first compare the train control performance H_2 norm in different schemes. Recall that the square root of the linear quadratic cost performance measure in Equation 7.4 is equivalent to the H_2 norm. We refer to the train control performance measure as the H_2 norm due to this equivalence. In this simulation scenario, the three trains depart from a station successively with interval $T_{\text{interval}} = 16\,\text{s}$ and the deceleration α that defines the ATP service braking curve is set to be $1.2\,\text{m/s}^2$. As we can see from Figure 7.6, the existing scheme gives the highest H_2 norm compared to the other three schemes. This is because the existing scheme does not use CoMP in the system, which can significantly improve train control performance. Moreover, it does not consider the dynamic transitions of wireless channels in CBTC systems, which is very important information for the train Station Adapter (SA) to make optimal handoff decisions to get the optimal performance.

By contrast, the other three schemes can have significant performance improvement due to the adoption of CoMP. The two SMDP-based schemes give better performance compared with the greedy scheme. This is because the SMDP model considers the dynamic transitions of wireless channels in CBTC systems. The direct optimization objective of the SMDP model is to minimize the linear quadratic cost of train control. Our proposed scheme gives the lowest H_2 norm. This is because the newly calculated guidance trajectory takes full consideration of the tracking

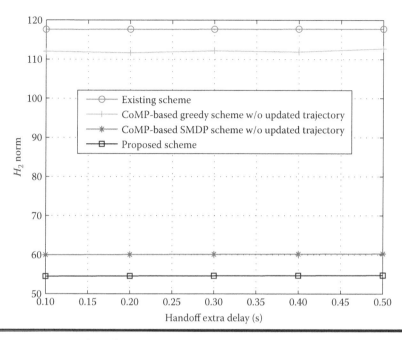

Figure 7.6 Control performance H_2 norm in different schemes.

error caused by handoff latency. As a result, it reduces the tracking error, which further improves the H_2 norm performance.

The train travel trajectories in our proposed scheme and the existing scheme are shown in Figures 7.7 and 7.8, respectively. The first, second, and third departed trains are denoted as Train 0, Train 1, and Train 2, respectively, in our simulations. We assume that the first departed train travels without obstacles because there are no trains traveling in front. As we can observe from the figure, compared with the existing scheme, the guidance trajectory in our proposed scheme for Train 1 and Train 2 is very close to Train 0. This is because the guidance trajectory is frequently recalculated to make up for the extra travel time caused by communication latency in this scheme. The figure also illustrates that, given the same train departure time interval, the travel trajectories of Train 1 and Train 2 in our proposed scheme are very close to the original optimized guidance trajectory. Although for the existing scheme, these two trains go through many unnecessary accelerations and decelerations during the train trip. This is because our scheme explicitly optimizes the train control performance. The service brake time caused by ATP subsystem is decreased to a minimum under the proposed optimal scheme, which helps mitigate the tracking errors.

Figure 7.9 shows the error between the real travel time and the preset trip time for Train 2 in different schemes. As shown in the figure, the travel time error is significantly increased when the trip time increases for the existing scheme and

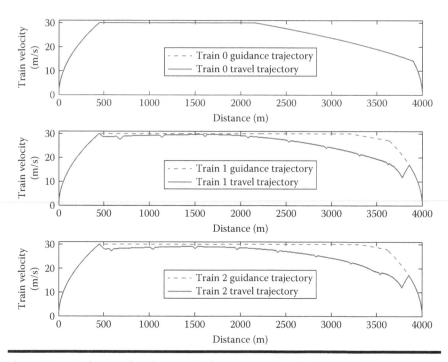

Figure 7.7   Train travel trajectory in the proposed CBTC system with CoMP.

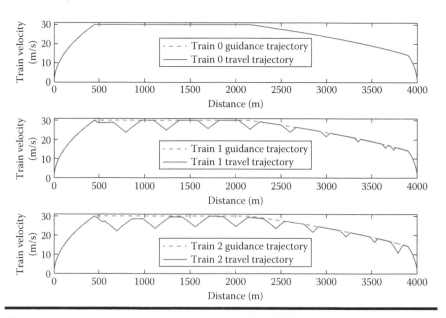

Figure 7.8   Train travel trajectory in the existing CBTC system.

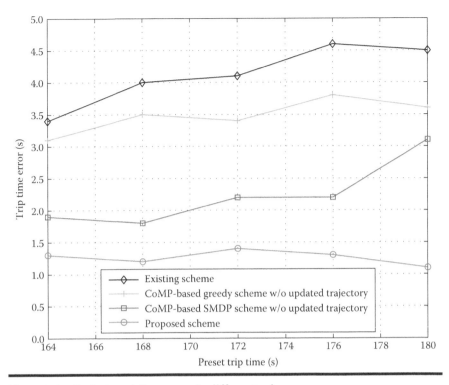

Figure 7.9 Train travel time error in different schemes.

the other two schemes. This travel time error severely affects the rail transit load capacity. By contrast, the train travel time in our proposed scheme is very close to the preset trip time. It successfully mitigates the wireless communication impacts on CBTC performance without decreasing the utilization of railway network infrastructure.

7.6.2 Handoff Performance Improvement

Figure 7.10 shows example handoff policies when three different schemes are used. In this figure, the Y axis represents the action. Action $= 1$ means the MT communicates with BS1 and CoMP is not used; Action $= 2$ means the MT communicates with BS2 and CoMP is not used; Action $= 3$ means the MT is working in the CoMP mode, and it communicates with the cluster where BS1 and BS2 are included.

As we can observe from the figure, there are only actions 1 and 3 in the existing CBTC scheme, where CoMP is not adopted. More importantly, there are more than one handoff events when the train travels across the boundary between two BSs, which is called ping-pong handoff. This is because the handoff decision

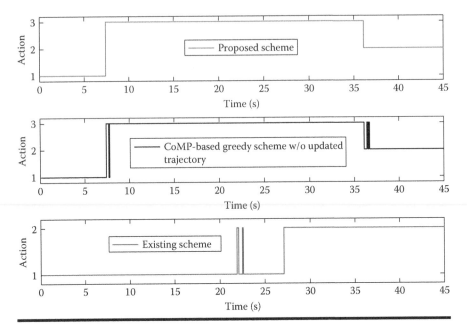

Figure 7.10 Handoff policies in different schemes.

policy in the existing scheme makes handoff decisions based on the current channel information and does not consider the dynamic transitions of wireless channels in CBTC systems, which is very important information to make handoff
decisions.

By contrast, three actions all appear in the schemes when CoMP is adopted.
The ping-pong handoff happens in the CoMP-based greedy scheme w/o updated
trajectory. This is because this scheme does not consider the dynamic transitions
of wireless channels in CBTC systems. In our proposed scheme, there is only a
small portion of the time when CoMP is not used. This is because the CoMP
mode performs better in most of the time. When the MT is close to a base station,
CoMP mode is not adopted. This is because the received signal is already good
enough under this circumstance. The lost and/or outdated channel state information due to constrained backhaul network will degrade the performance. We can
also observe that a handoff is performed by switching action from 1 to 3, and then
switching from 3 to 2. There are only two action switches. In other words, there is
no ping-pong handoff in the proposed scheme. This is because the SMDP model is
used in the proposed scheme to calculate the handoff decision policy. This model
takes full consideration of the dynamic transitions of wireless channels in CBTC
systems.

Following [32], we take service discontinuity as another performance metric to
compare the proposed scheme with other schemes. Figure 7.11 shows the average
service discontinuity time duration in different schemes. We can observe that the

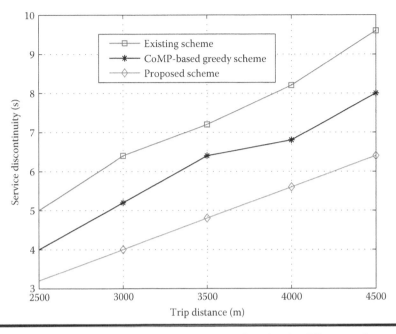

Figure 7.11 **Average service discontinuity time duration in different schemes.**

existing scheme gives the highest average service discontinuity time duration compared to the other schemes. This is because the CoMP is not used in this scheme. Our proposed scheme gives better average service discontinuity time duration compared with the CoMP-based greedy scheme. This is because CoMP improves service availability and no ping-pong handoff happens in the CBTC system under the proposed scheme, which consequently decreases the average service discontinuity time duration.

7.7 Conclusion

Train–ground communication is one of the key technologies in CBTC systems. Unreliable wireless communications and frequent handoffs have significant impacts on the train control performance in CBTC systems. In this chapter, using recent advances in CoMP, we proposed a novel CBTC system, in which a train can communicate with a cluster of base stations simultaneously to enhance the train control performance of CBTC systems. In order to mitigate the impacts of wireless communications on CBTC performance, we took a cross-layer design approach to optimize the control performance in CBTC systems. Unlike the existing works on CoMP that use traditional design criteria, such as network capacity, we used linear quadratic cost for the train controller as the performance measure.

We jointly considered base station cluster selection and handoff decision issues in CBTC systems. An optimal guidance trajectory calculation scheme was proposed in the train control procedure to further improve the performance of CBTC systems. Simulation results were presented to show that the proposed approach can significantly improve the train control performance and increase the railway capacity.

References

1. J. Huang, J. Ma, and Z. Zhong. Research on handover of GSM-R network under high-speed scenarios. *Railway Commun. Signals*, 42:51–53, 2006.
2. C. Yang, L. Lu, C. Di, and X. Fang. An on-vehicle dual-antenna handover scheme for high-speed railway distributed antenna system. In *Proc. IWCMC'10*, Caen, France, 2010.
3. Z. Li, F. R. Yu, B. Ning, and T. Tang. Cross-layer handoff design in MIMO-enabled WLANs for communication-based train control (CBTC) systems. *IEEE J. Sel. Areas Commun.*, 30(4):719–728, 2012.
4. R. Irmer, H. Droste, P. Marsch, M. Grieger, G. Fettweis, S. Brueck, H.-P. Mayer et al. Coordinated multipoint: Concepts, performance, and field trial results. *IEEE Comm. Mag.*, 49(2):102–111, 2011.
5. J. Tang and X. Zhang. Cross-layer modeling for quality of service guarantees over wireless links. *IEEE Trans. Wireless Commun.*, 6(12):4504–4512, 2007.
6. J. Tang and X. Zhang. Cross-layer resource allocation over wireless relay networks for quality of service provisioning. *IEEE J. Sel. Areas Commun.*, 25(4):645–657, 2007.
7. A. L. Toledo and X. Wang. TCP performance over wireless MIMO channels with ARQ and packet combining. *IEEE Trans. Mobile Comput.*, 5(3):208–223, 2006.
8. F. R. Yu, B. Sun, V. Krishnamurthy, and S. Ali. Application layer QoS optimization for multimedia transmission over cognitive radio networks. *ACM/Springer Wireless Networks*, 17(2):371–383, 2011.
9. P. Dorato, V. Cerone, and C. Abdallah. *Linear-Quadratic Control: An Introduction.* Simon & Schuster, New York, 1990.
10. H. Dong, B. Ning, B. Cai, and Z. Hou. Automatic train control system development and simulation for high-speed railways. *IEEE Circuits and Systems Magazine*, 10(2):6–18, 2010.
11. M. Puterman. *Markov Decision Processes: Discrete Stochastic Dynamic Programming.* John Wiley & Sons, New York, 1994.
12. Q. Zhao, L. Tong, A. Swami, and Y. Chen. Decentralized cognitive MAC for opportunistic spectrum access in ad hoc networks: A POMDP framework. *IEEE J. Sel. Areas Commun.*, 25(5):589–600, 2007.
13. P. G. Howlett, P. J. Pudney, and X. Vu. Local energy minimization in optimal train control. *Automatica*, 45(11):2692–2698, 2009.
14. P. Howlett. The optimal control of a train. *Ann. Oper. Res.*, 98(1–4):65–87, 2000.
15. Q. Song and Y.-D. Song. Data-based fault-tolerant control of high-speed trains with traction/braking notch nonlinearities and actuator failures. *IEEE Trans. Neural Net.*, 22(12):2250–2261, 2011.
16. F. Babich, G. Lombardi, and E. Valentinuzzi. Variable order Markov modeling for LEO mobile satellite channels. *Electron. Lett.*, 35(8):621–623, 1999.

17. F. Babich and G. Lombardi. A measurement based Markov model for the indoor propagation channel. In *Proc. IEEE VTC*, volume 1, Phoenix, AZ, May 1997.
18. H. S. Wang and P.-C. Chang. On verifying the first-order Markovian assumption for a Rayleigh fading channel model. *IEEE Trans. Veh. Tech.*, 45(2):353–357, 1996.
19. H. S. Wang and N. Moayeri. Finite-state Markov channel—A useful model for radio communication channels. *IEEE Trans. Veh. Tech.*, 44(1):163–171, 1995.
20. C. Pimentel, T. H. Falk, and L. Lisbôa. Finite-state Markov modeling of correlated Rician-fading channels. *IEEE Trans. Veh. Tech.*, 53(5):1491–1501, 2004.
21. Y. L. Guan and L.F. Turner. Generalized FSMC model for radio channels with correlated fading. *IEE Proc. Commun.*, 146(2):133–137, 1999.
22. C. D. Iskander and P. T. Mathiopoulos. Fast simulation of diversity Nakagami fading channels using finite-state Markov models. *IEEE Trans. Broadcasting*, 49(3):269–277, 2003.
23. J. Yang, A. K. Khandani, and N. Tin. Statistical decision making in adaptive modulation and coding for 3G wireless systems. *IEEE Trans. Veh. Tech.*, 54(6):2066–2073, 2005.
24. L.-C. Wang and C.-J. Yeh. 3-Cell network MIMO architectures with sectorization and fractional frequency reuse. *IEEE J. Sel. Areas Commun.*, 29(6):1185–1199, 2011.
25. D. Gesbert, S. Hanly, H. Huang, S. Shamai Shitz, O. Simeone, and W. Yu. Multi-cell MIMO cooperative networks: A new look at interference. *IEEE J. Sel. Areas Commun.*, 28(9):1380–1408, 2010.
26. K. Woradit, T. Q. S. Quek, W. Suwansantisuk, H. Wymeersch, L. Wuttisittikulkij, and M. Z. Win. Outage behavior of selective relaying schemes. *IEEE Trans. Wireless Commun.*, 8(8):3890–3895, 2009.
27. S. Lin, D. Costello, and M. Miller. Automatic-repeat-request error-control schemes. *IEEE Comm. Mag.*, 22(12):5–17, 1984.
28. C. Rico-Garcia and O. K. Tonguz. Increasing safety and efficiency of railway. In *Proc. IEEE ICC'12 Workshop on Intelligent Vehicular Networking: V2V/V2I Comm. App.*, Ottawa, Canada, June 2012.
29. H. J. Kushner and G. G. Yin. *Stochastic Approximation and Optimization of Random Systems*. Springer, New York, 2003.
30. G. H. Lee, I. P. Milroy, and A. Tyler. Application of Pontryagin's maximum principle to the semi-automatic control of rail vehicles. In *Proc. 2nd Conference on Control Engineering*, Newcastle, Australia, 1982.
31. L. Zhu, F. R. Yu, and B. Ning. A seamless handoff scheme for train-ground communication systems in CBTC. In *Proc. IEEE VTC*, Ottawa, Canada, September 2010.
32. J.-M. Chung, M. Kim, Y.-S. Park, M. Choi, S. Lee, and H. S. Oh. Time coordinated V2I communications and handover for WAVE networks. *IEEE J. Sel. Areas Commun.*, 29(3):545–558, 2011.

Chapter 8

Novel Handoff Scheme with Multiple-Input and Multiple-Output for Communications-Based Train Control Systems

Hailin Jiang, Victor C. M. Leung,
Chunhai Gao, and Tao Tang

Contents

8.1 Introduction 150
8.2 Overview of the CBTC Communication Handoff Procedure 152
    8.2.1 Handoff Latency of 802.11 and the Communication Latency
        in CBTC Systems 152
    8.2.2 Features of Handoff in CBTC Communication System 154
8.3 Proposed MAHO Scheme 154
    8.3.1 MIMO Transmission in the Handoff Procedure
        in the MAHO Scheme 155
        8.3.1.1 Physical Layer Processing 155
        8.3.1.2 Synchronization in the Downlink 159
    8.3.2 Communication Latency in WLANs 161
8.4 Analysis of Handoff Performance 162
    8.4.1 Wireless Channel Model 162

8.4.2 Optimal Handoff Location..163
8.4.3 Error-Free Period 164
8.4.4 FER of the Handoff Signaling 166
8.4.5 Impacts on Ongoing Data Sessions..167
8.5 Simulation Results and Discussions 167
8.5.1 Analysis of the Handoff Latency..167
8.5.2 Error-Free Periods of Traditional Handoff Schemes....................169
8.5.3 FER of Handoff Signaling with Different Data Rates..................170
8.6 Conclusion..173
References 174

8.1 Introduction

The IEEE 802.11 standard for wireless local area networks (WLANs) has been developed to provide wireless access in the office and campus environments. However, the procedures for handoff between access points (APs) are not well supported. When a train moves along the rail, its mobile station (MS) would need to switch from one AP to the next frequently to guarantee continuous data transmissions between the train and the wayside devices, because the coverage of each AP is quite limited. In general, during the handoff procedure data packets will be lost if no AP forwarding schemes are implemented. This will have serious impacts on the safety and efficiency of train control.

A lot of research has been done on the WLAN handoff algorithms. The handoff procedure is divided into three stages [1]: the probe stage, the searching stage, and the executing stage. In [1], some 802.11b parameters are further adjusted to reduce the handoff interruption time. In the current version of the IEEE 802.11 standards [2], the formats of probe request and probe response frames are defined and the fast basic service set transition schemes including authentication and reassociation schemes are given. In [3], a channel scanning scheme in WLAN handoff is proposed, where the neighbor cells were cached in the buffer to reduce the handoff latency in the probe stage. It is proposed in [4] that each MS continuously tracks nearby APs by synchronizing short listening periods with periodic transmissions from each base station. In this way, the station can pre-associate with new APs to reduce the handoff latency. A location-based handoff scheme is proposed in [5], and some configurations of the parameters in 802.11 networks are discussed to reduce the handoff latency in the scanning stage. An integrated design approach is proposed to jointly optimize handoff decisions and physical layer parameters to improve the train control performance in CBTC WLAN systems in [6–8].

These schemes can reduce the handoff latency efficiently, but all of them are so-called break-before-make schemes, where the station needs to dissociate with the old AP first, and then find a new AP and associate with it. In this procedure, the data transmissions will be interrupted and the transmitted data will be lost. In a CBTC system employing WLAN technology, the data transmitted include safety-related train control and train position information, and the loss of the data will lower the performance of CBTC control severely and affect the system safety.

In [9–11], multiple radio transceivers are used to eliminate the handoff latency completely. However, these schemes are required to use two or more radio units to scan and pre-associate with new APs, which will increase the cost and complexity of the communication system.

Multiple-input and multiple-output (MIMO) has been the key technology in mainstream wireless communication technical standards such as 3rd Generation Partnership Project (3GPP) Long-Term Evolution (LTE) and IEEE 802.11n [12]. Some handoff schemes applying MIMO technologies are proposed in [13,14]. In [13], the handover decision method based on detecting the number of antennas is proposed. In [14], the beamforming and positioning-assisted handover scheme is proposed where both the source eNodeB and the target eNodeB switch their working mode from omnidirectional to beamforming to improve the handover success probability when the train moves into the overlapping region.

In this chapter, we propose a MIMO-assisted handoff (MAHO) scheme for CBTC WLAN system with two antennas or more antennas con–d on train and each AP. The distinct features of the work are given as follows:

1. Considering the features of the urban railway communication system, the triggering of the handoff is based on the location of the train. In the scheme, the MS does not hand off to the new AP by comparing the signal strength, signal-to-noise ratio (SNR), or any other signal transmission-related parameters of the APs. Instead, it will start the handoff procedure when the train has reached the handoff location. The location-based handoff is proposed to take advantage of the fact that the train in the CBTC system can acquire its locations accurately in real time.
2. The train position information from the MS is sent simultaneously with handoff signaling by means of MIMO multiplexing. The signaling and data packets are recovered at the APs by means of MIMO signal detection algorithms such as vertical-bell laboratories layered space-time (V-BLAST) [15] algorithms.
3. The train control information from different APs is sent in the space–time block code (STBC) diversity mode, where simple linear interference canceling algorithms are used to cancel the interference caused by the concurrent transmissions.
4. The handoff performance, including the frame error rate (FER) of the handoff signaling, the handoff latency, the error free period, and the impacts on the intersite distance, is analyzed and compared with traditional handoff schemes.

In the proposed scheme, the handoff procedure is triggered by the location of the train. When the handoff is triggered, the station keeps connection with old AP by one antenna on the train and set up a connection with new AP by another antenna at the same time. The interference at the station between two APs is canceled by MIMO signal detection techniques and interference canceling algorithms. In CBTC environments, the APs are planned to assure that even in the border of coverage area the received power exceeds the minimum receiver sensitivity by more than 10 dB [5], which guarantees the reliability of the data transmission and makes the signal detection algorithm feasible. With the SNR over certain thresholds, the MIMO signal detection and interference cancelation algorithms can be reliably applied and the "make-with-break" can be realized without multiple radio transceivers at the MS.

The rest of the chapter is organized as follows: An overview of handoff procedure in CBTC systems is given in Section 8.2. The MAHO scheme is presented and the synchronization and interference canceling algorithms are described in Section 8.3. The handoff performance is analyzed and compared with existing schemes in Section 8.4. Simulation results are presented in Section 8.5. Conclusion is drawn in Section 8.6.

8.2 Overview of the CBTC Communication Handoff Procedure

8.2.1 Handoff Latency of 802.11 and the Communication Latency in CBTC Systems

The timing diagram of the handoff procedures in 802.11 is shown in Figure 8.1. There are three stages in the handoff procedure: the probe, authentication, and reassociation procedures. In passive scanning mode, the MS receives the periodic Beacon frame from APs in the probe stage, then finds the suitable AP to set up the communication link. The transmission period of the Beacon frame is about 100 ms, so the latency in this stage is several hundreds of milliseconds, which is too long for the CBTC service. In the active scanning mode, in general, the MS broadcasts the probe request at all channels one by one, then waits for the probe response

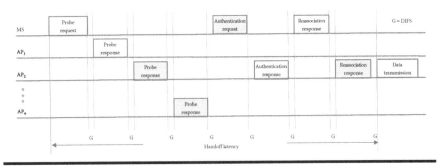

Figure 8.1 WLAN handoff timing diagram.

from APs. As there are 11 channels defined in 802.11 [2], the latency in this stage is quite long. Fortunately, only one channel is used in CBTC communication system for all APs along the rails. The MS on the train is generally configured to only scan one channel to reduce the scanning latency.

At the authentication stage, the MS authenticates with the best AP found in the first stage. When the 802.1X authentication is implemented in the WLAN, the latency is more than 150 ms [16], which is too expensive for the CBTC communication system. For the authentication procedure of over-the-air fast transition protocol between APs with four-way handshake defined in 802.11 standard [2], the latency in this stage is about 40 ms [16].

CBTC systems have stringent requirements for communication latency. The data between the train and the ZC are transmitted periodically. Because a train needs to get the location information of the train ahead of it, trains and ground equipment must communicate with each other in every communication period.

When the communication latency is short, a train gets the position of the train ahead of it in every communication period, and the minimum train headway (minimum distance between two successive trains) is calculated considering the braking distance, the constant safe distance, and the train length. When the communication latency is long, a train may not be able to get the position of the train ahead of it at some communication periods. Therefore, the train has to suppose that the preceding train is still at the position which was got from the previous communication period. The consequence is that the train has to brake in advance, which will affect the utilization seriously.

When a train moves between successive APs, the channels between the train and the wayside APs change rapidly due to the fast movement of the train. And this will result in rapid changes of the received power and signal quality. The rapid changes of the communication link also have serious effects on the frequent handoff procedures, which may result in long transmission latency.

In global system for mobile communication/railway (GSM-R) communication system, there are two special quality-of-service (QoS) targets to fill for the train's operation requirements [18,17]. The first is the transmission interference period, which is the period during the data transmission phase of an existing connection in which no error-free transmission of user data is possible. All the packets will be lost during the handoff procedures if there are no forwarding schemes between APs. It is clear that the transmission interruption period is just the handoff latency in CBTC communication system because the SNR margin is generally set to be large enough that packets will not be lost in normal transmission process. And the interruption due to handoff procedures will result in the increase of the end-to-end communication latency in CBTC communication system.

The other parameter is error-free period, which follows the transmission interference period to retransmit user data in error and data units waiting to be served. There are no standards or drafts for CBTC communications in the urban railway transition system. However, it is clear that the transmission interruption period, just the handoff latency in this chapter, should be as small as possible and the error-free period should be as large as possible. These two QoS indexes, together with the FER of the handoff signalings, will be used to evaluate the QoS performance in CBTC communication system in this chapter.

8.2.2 Features of Handoff in CBTC Communication System

The APs in the CBTC system in urban rail transit system are located with a small distance of several hundred meters between each other. As the APs are arranged along the rail in fixed positions, the mobile station in the train would switch from one AP to another predetermined AP. Therefore, the handoff algorithm can be simplified and optimized to avoid ping-pong handoff and inappropriate handoff.

Another distinct feature of CBTC system is that the train can acquire high-resolution train location determination [19] by speed sensors and Doppler radar together with the transponders, which are aside the rail and provide the absolute reference position information for the train. The train transmits the location information to the wayside device by the wireless communication system. According to the requirements of the CBTC system, the location resolution would be in the limit of 5–10 m. Because the location of the train can be got in real time, we propose the location-based handoff algorithm in [5] to reduce the happening of ping-pong handoff caused by the random effects of the wireless channel.

In CBTC systems, it is important to maintain communication link availability in order to guarantee train operation safety and efficiency. In this chapter, on the basis of the location-based handoff scheme, we present a "make-with-break" handoff scheme to provide high link availability in CBTC systems by applying MIMO multiplexing and STBC interference cancelation algorithms.

8.3 Proposed MAHO Scheme

Two or four antennas can be configured in each AP and the MS on the train according to [12]. We suppose that the MAHO scheme with 2×2 antenna configuration be applied in the chapter, and the method can be easily expanded to the case of four antennas as defined in 802.11n standard [12].

In the proposed MAHO scheme, the handoff procedure is triggered by the location of the train. When the handoff is triggered, the station would keep connection with old AP by one antenna on the train and set up a connection with the new AP by another antenna at the same time. The interference at the MS between two APs is canceled by MIMO signal detection and interference canceling algorithms. In CBTC environments, the wireless networks are planned to assure that even in the border of coverage area the received power exceeds the minimum receiver sensitivity by more than 10 dB [5], which guarantees the reliability of the data transmission and make the signal detection algorithm feasible. With the strong enough

SNR, the MIMO signal detection and interference cancelation algorithms can be reliably applied and the "make-with-break" scheme is realized without multiple radio transceivers.

8.3.1 MIMO Transmission in the Handoff Procedure in the MAHO Scheme

When the train reaches the handoff point, the MS will initiate the handoff procedure by transmitting the probe request packet. The APs receiving the request will send probe response to the MS. In our proposed MAHO scheme, the candidate AP is predetermined along the rail, and the MS can decode the probe response signal from the candidate AP. The MS can learn whether the candidate works well by receiving the periodic Beacon frame from the candidate AP.

According to the regulations in the standards, the probe response packets have to be transmitted in sequence and the delay in this stage is the largest part of the handoff latency. In the MAHO scheme, handoff signaling and uplink data are transmitted in MIMO multiplexing mode in the uplink, whereas in the downlink the multiuser STBC transmission modes are used to reduce the latency.

The physical layer-related topics are first introduced in Section 8.3.1.1, then the related synchronization schemes are given.

8.3.1.1 Physical Layer Processing

As shown in Figure 8.2a, if at the time instant of sending the probe request there is an uplink train control-related information to send, the MS sends the uplink (UL) data together with probe request, where the handoff signaling and the data packet are transmitted by different antennas, respectively. And the decoding of the MIMO multiplexing packets can apply the well-known V-BLAST algorithm [15]. In such a way, the data communication will not be interrupted by the handoff procedure. When the serving AP receives the UL data multiplexed with the probe request packet, it returns the ACK packet sending at one antenna and the probe response at another antenna, which are transmitted in MIMO multiplexing mode. The packets are transmitted after the period of short interframe space (SIFS), which is required before sending the ACK packet as defined in 802.11 standard.

However, the candidate AP senses the channel status and returns a probe response after the period of distributed interframe space (DIFS), plus the period that the contention window (CW) timer expires if the channel is idle, where the transmission mode is the MIMO diversity by using Alamouti STBC [20].

The MIMO transmission mode of the transmission is indicated in high throughput signal field part of the frame header in 802.11 packets, where the number of spacial streams, N_{SS}, and space time streams, N_{STS}, are included. For the MIMO multiplexing mode, $N_{SS} = 2$ and $N_{STS} = 2$, and for the MIMO STBCs, $N_{SS} = 1$ and $N_{STS} = 2$.

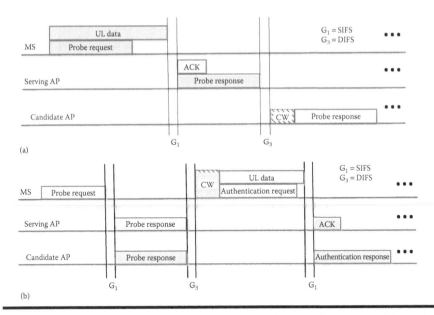

Figure 8.2 Proposed handoff scheme. (a) Multiplexing of UL data with probe request. (b) Synchronized downlink transmission.

If there are no uplink train control-related data to send at the time instant of sending the probe request as shown in Figure 8.2b, concurrent transmission of the probe response packets in the downlink is implemented to reduce the latency in the scanning stage.

After the transmission of probe response packets, the MS will transmit other handoff signaling packets together with UL data packets in the MIMO multiplexing mode if there are UL packets to send at that same time. The serving AP and the candidate AP will return the ACK packet for the UL data packet and the handoff signaling packets simultaneously as shown in Figure 8.2b, which is the same as the transmission of probe response packets. Otherwise, if there are no data packets to send at the time of transmitting the handoff signaling packets, the signaling packets are transmitted at both antennas in the MIMO diversity mode.

In our transmission scheme, both APs use Alamouti code and the packets are transmitted to the MS simultaneously, which will cause interference between the two APs at the receiver. However, the interference can be canceled with simple linear algorithms [21,22]. Let $\mathbf{c} = (c_1\ c_2)^T$ and $\mathbf{s} = (s_1\ s_2)^T$ be the codewords transmitted by the first and second APs, respectively. $\mathbf{y}_j = (y_{1j}\ y_{2j}^\star)^T$ is the received signal vector at the antenna j, where the components of $\mathbf{y}_j$ are the signals received at the antenna j over two consecutive symbol periods. Therefore, we have

$$
\begin{aligned}
\mathbf{y}_j &= \begin{pmatrix} y_{1,j} \\ y_{2,j}^{\star} \end{pmatrix} = \begin{pmatrix} h_{1,j} & h_{2,j} \\ h_{2,j}^{\star} & -h_{1,j}^{\star} \end{pmatrix} \cdot \begin{pmatrix} c_1 \\ c_2 \end{pmatrix} \\
&+ \begin{pmatrix} g_{1,j} & g_{2,j} \\ g_{2,j}^{\star} & -g_{1,j}^{\star} \end{pmatrix} \cdot \begin{pmatrix} s_1 \\ s_2 \end{pmatrix} + \begin{pmatrix} n_1 \\ n_2 \end{pmatrix}
\end{aligned}
\tag{8.1}
$$

where:

$h_{i,j}$ is the channel coefficient from the antenna i of AP_1 to the j receive antenna of MS

$g_{i,j}$ is the channel coefficient from the antenna i of AP_2 to the j receive antenna of MS

The noise samples n_1, n_2 are independent samples of a zero-mean complex Gaussian random variable

And we define the channel matrix as

$$
\mathbf{H}_j = \begin{pmatrix} h_{1,j} & h_{2,j} \\ h_{2,j}^{\star} & -h_{1,j}^{\star} \end{pmatrix}, \mathbf{G}_j = \begin{pmatrix} g_{1,j} & g_{2,j} \\ g_{2,j}^{\star} & -g_{1,j}^{\star} \end{pmatrix}
$$

A simple array signal processing results in the following signal:

$$
\begin{aligned}
&\frac{\mathbf{G}_2^H \mathbf{y}_2}{|\mathbf{G}_2|^2} - \frac{\mathbf{G}_1^H \mathbf{y}_1}{|\mathbf{G}_1|^2} \\
&= \left(\frac{\mathbf{G}_2^H \mathbf{G}_1}{|\mathbf{G}_2|^2} - \frac{\mathbf{H}_2^H \mathbf{H}_1}{|\mathbf{H}_2|^2} \right) \begin{pmatrix} c_1 \\ c_2 \end{pmatrix} + \begin{pmatrix} n_1' \\ n_2' \end{pmatrix}
\end{aligned}
\tag{8.2}
$$

In the above equation, the signal of the AP_2 has been canceled, and the decoding of the signal of AP_1 is straightforward. This array processing technique provides a diversity of degree 2 for each AP.

In the case that the train and APs have four antennas each, multiplexing of handoff signaling with data packets can be implemented in the uplink, where some antennas are used to transmit signaling and the others are used for the transmission of data packets. The decoding of the multiplexed packets can apply the V-BLAST algorithm [15] as well. In the downlink that APs transmit packets to the train, because orthogonal STBC for complex signal transmission does not exist [23], a quasiorthogonal STBC (QOSTBC) transmission for the AP in the downlink can be implemented, where the MS on the train can cancel the interference of other APs through array processing [24].

In the proposed MAHO scheme, the handoff signaling and train control-related data packets are transmitted at different antennas to reduce the handoff latency. However, the concurrent transmission of data and signaling packet will cause more packet loss compared with the transmission of signaling or data packets by transmit diversity mode, even with the interference cancelation algorithm employed. The required data rate in CBTC communication system is less than 1 Mbps [25]; therefore, the modulation and coding scheme (MCS) of 6.5 Mbps in 802.11n is commonly applied, which is also the most robust transmission scheme in 802.11n system with orthogonal frequency division multiplexing (OFDM) modulation. The moving speed of the train in the subway is not more than 80 km/h in general, so the maximum Doppler frequency shift of the 2.4 GHz WLAN system is at most 200 Hz. Although the subcarrier bandwidth in 802.11g system is 312.5 KHz, the normalized frequency offset in the OFDM system is about 6.4×10^{-4}, which has nearly no impacts on the BER performance of the system [26]. In [27], it is proposed that when the zero forcing algorithm is adopted at the receiver, the BER is approximated as

$$P_b \approx \alpha \gamma^{-\beta} \tag{8.3}$$

where:
γ is the average received SNR
α and β are related with the achieved channel degrees of freedom (DoF)

With 6.5 Mbps MCS in CBTC communication system, where binary phase shift keying (BPSK) modulation with coding rate of 1/2 is adopted, the DoF is 2, and α is 0.0272 and β is 7.528 when MIMO multiplexing mode is used with two streams multiplexed, which is the case of the uplink transmission mode in our scheme. In the scheme of concurrent transmission in the downlink by two APs, 2×2 Alamouti STBC codes are used at both APs and two antennas at the receiver at the MS and the DoF is 4. The corresponding coefficients of α and β are 0.0031 and 9.853, respectively. When 2×2 Alamouti STBC codes are adopted in single user's transmission, which is the case of normal data transmission with transmit diversity and receiver diversity, the values are 0.0008 and 9.540, respectively. These give an upper bound of the BER of the packet transmission in the proposed handoff MAHO scheme.

Then the BER results are used to calculate the FER P_{fr}:

$$P_{fr} = 1 - (1 - P_b)^{L_{fr}} \tag{8.4}$$

where:
P_b is the BER

The curves of FER P_{fr} versus E_s/N_0, with different data rates and channel DoF are given in Figure 8.3 with the frame length of 150 bytes.

To apply the interference cancellation algorithm, the packets from different APs have to arrive at the MS at the same time instant and synchronized at the receiver. The synchronization methods are introduced in Section 8.3.1.2.

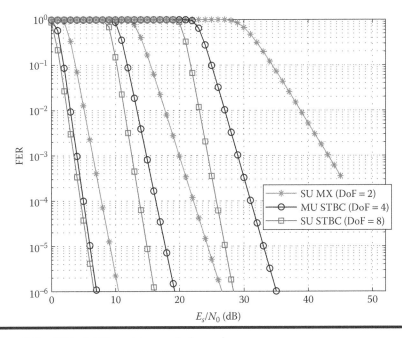

Figure 8.3   FER of different transmission schemes.

8.3.1.2 Synchronization in the Downlink

Each AP receiving the probe request packet will return a probe response to the MS in sequence in the existing scheme, which spends a big part of the time in the handoff procedures. And the MS may receive a lot of probe response signals from the rail-side APs and other interference sources in this stage. However, in CBTC environments only the packets from the serving AP and the handoff candidate AP are the expected ones. To avoid the unnecessary transmission of the probe response packet, the MS transmits the location information in the "Vendor Specific" domain of probe request packet. Thus, the APs learn the location of the MS and will send the probe response packets only when having received the probe request packet from the specific handoff locations, which ensures that only the two involved APs return the probe response packets. And the other APs outside the CBTC WLAN will not respond to the probe request packet because the service set identifier (SSID) of the CBTC WLAN is unknown to them.

To reduce the latency in this stage, we propose that the serving AP and the handoff candidate AP send the probe response packets synchronously in the downlink. However, in WLAN system the packets are transmitted with carrier sense multiple access with collision avoidance (CSMA/CA) protocol, and the random backoff will happen before the transmission of probe response. If the packets sent by the different APs have different backoff delay, the packets cannot arrive at the receiver at the same time even though they are transmitted simultaneously with the transmission delay considered. To achieve the synchronous reception of the packets, revisions of the packet transmission scheme in 802.11 standards are implemented. Therefore, we set that the two APs receiving the probe request transmit the probe response packets after the SIFS interval, instead of the DIFS plus the backoff duration defined in the specification. In this way, the arrival time at the MS is synchronized and the interference cancelation algorithm can be implemented. However, the application of SIFS also guarantees that the probe response packets from unwished sources cannot be sent earlier than the expected packets, and the timer of the MaxChannelTime, which is the amount of time to wait to collect potential additional probe responses from other access points, can be set to small enough to reduce the latency in the scanning stage. The SNRs of the signals from the serving AP and the handoff candidate AP are much stronger than the interference signals in general; therefore, the subsequent transmissions of the probe response packets have little impacts on the communication performance.

In the proposed scheme, the handoff triggering location is predefined and strong line-of-sight (LOS) components exist between the AP and the MS, so the propagation delay between the MS and the two APs can be got in advance. For the APs involved in the handoff procedures, there are four handoff locations as shown in Figure 8.4, where two are on either side of AP_2. Then the propagation delay $tp_{i,j}$ for the AP_2 is calculated as

$$tp_{i,j} = d_{i,j}/c, \quad i = 1,\ldots,4, \quad j = 1,3 \tag{8.5}$$

where:

c is the speed of light

$d_{i,j}$ is the distance from the MS to the AP_j at the ith handoff location

The delay differences of the four handoff locations around the AP_2 are calculated as

$$\Delta t_{i,2} = tp_{i,2} - tp_{i,j}, \quad i = \{1,3\}, j = 1; i = \{2,4\}, j = 3 \tag{8.6}$$

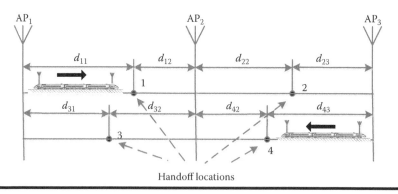

Handoff locations

Figure 8.4   Delay difference between adjacent APs.

The AP_2 stores the delay difference of the handoff locations, and when the MS sends the probe request to the AP, the AP learns the position and the moving direction of the MS from the received packets, and then knows that it is at the handoff location point i. Then the AP can transmit the probe response packet to the MS, with the delay difference Δt_i counted in, to guarantee the synchronous reception at the MS. Suppose the handoff from AP_1 to AP_2 initiates at handoff location 1, and the time instants that the AP_1 and AP_2 receiving the probe request are T_{rx1} and T_{rx2}, respectively, and the processing delay and transmission delay are $t_{processing}$ and t_{tx}, which are supposed the same for all the APs in this chapter. Without loss of generality we suppose the transmission delay at the first handoff location has $\Delta t_{1,2} < 0$, then the transmit instants for the AP_1 and AP_2 to send the probe response packet are calculated as follows:

$$T_{tx1} = T_{rx1} + t_{processing} + t_{tx} \tag{8.7}$$

$$T_{tx2} = T_{rx2} + t_{processing} + t_{tx} - 2 \times \Delta t_{1,2} \tag{8.8}$$

From the equations, we can see that the APs only have to store the delay differences with negative values. One example is with the format of ($\Delta t_{1,j}$, $\Delta t_{4,j}$).

In the CBTC communication system, OFDM is adopted due to its anti-multipath characteristics, with the cyclic prefix (CP) length of 0.8 μs. With the CP, if the timing differences between uplink clients are in the limit of half of the CP length, 0.4 μs, the signal can be still be decoded successfully [28]. The accuracy of the location in CBTC system is about 5–10 m; therefore, the delay uncertainty between the two APs due to locating uncertainty is less than 0.067 μs, which is far less than the 0.4 μs and CP length of OFDM packets. Therefore, little uncertainty of the locating results has no effects on synchronization performance.

8.3.2 Communication Latency in WLANs

In the WLAN system in CBTC, CSMA/CA is the media access control (MAC) protocol for all the stations. Two reasons resulting in the loss of packets are as follows: (1) the loss because of the collisions when the different nodes send the packets simultaneously, which is caused by the contention of different transmitters, and (2) the transmission error, where a packet is received without packet collisions and is corrupted due to low SNR. In this chapter, we only consider the latter case because in typical CBTC environments there is only one train in the same direction in the AP's coverage range, which is decided by the requirements of the safety distance between consecutive trains.

With packet loss caused only by transmission error, the packet delay T_{ho_sig} of one single handoff signaling packet in CSMA/CA systems is calculated as follows:

$$T_{ho\_sig} = T_P + T_{tx} + T_{DIFS} + T_{CCA} + T_{RxTx}$$
$$+ T_{Preamble} + T_{PLCP} + N_{backof} \times T_{slot} \tag{8.9}$$

where:

T_p is the propagation latency

T_{tx} is the transmission latency

T_{DIFS} is the DIFS

T_{CCA} is the time to sense the channel

T_{RxTx} is the latency when the station transits from receiving to transmitting status

$T_{Preamble}$ and T_{PLCP} are the times to transmit preamble and physical layer convergence protocol (PLCP) packet, respectively

N_{backof} is the number of backoff slots before transmitting data

T_{slot} is the slot length in 802.11

T_{tx} can get as $T_{tx} = L_{fr}/R$, where R is the data rate and L_{fr} is the frame length in bits.

In normal handoff procedure, the handoff is divided into scanning, authentication, and reassociation stages, where in the scanning and reassociation stages, there are two signaling packets' transmission, and in the authentication stage, there are four signaling packets [2] in the fast transition protocol. Suppose that the sizes of all the handoff signaling are the same, and during the handoff procedure, the transmission error happens at most once due to the fact that the probability that two successive signaling packets both fail is very small. If any one of the authentication and reassociation packets is lost, the handoff procedure will start from the beginning of the handoff procedure. It will take some time before the SA starts over again. We refer to that time as T_{wait}. The handoff latency is calculated as

$$T_{HO\_latency} = (1 - P_{fr})^8 \times (T_{ho\_sig} + mChTime)$$
$$+ P_{fr}(1 - P_{fr})^7 \times \left[(T_{ho\_sig} + N_{backoff} \times T_{slot}) \times (0 + 1 + \ldots + 7) + mChTime + T_{wait} \right] \quad (8.10)$$

where:

$mChTime$ is the maximum time to spend on each channel when finding that the channel is not idle during the scanning stage

P_{fr} is the FER

8.4 Analysis of Handoff Performance

8.4.1 Wireless Channel Model

The transmitted signal will experience different fading of three factors: path loss, shadow fading, and fast fading. Fast fading such as Rician or Rayleigh fading has shorter correlation distance and faster variation compared with shadow fading, and the effects are usually averaged out when applying the trigger conditions of handoff. The path loss model for 2.4 GHz signal near the ground is proposed in [29] as follows:

$$P_{loss}(d) = 7.6 + 40\log_{10}d - 20\log_{10}h_t h_r \quad (8.11)$$

where:

d is the distance between the transmitter and the receiver

h_t and h_r are the height of the transmit antenna and receive antenna, respectively

Therefore, the received signal power from the serving AP_1 and candidate AP_2, expressed in dBm, when the distance between the MS and AP_1 is d, is formulated as follows:

$$a(d) = P_t + G_t + G_r - L_t - L_r - P_{loss}(d) + u(d) \tag{8.12}$$

$$b(d) = P_t + G_t + G_r - L_t - L_r - P_{loss}(D - d) + v(d) \tag{8.13}$$

where:

P_t is the transmit power

G_t and G_r are the transmitter and receiver antenna gains, respectively

L_t and L_r are the insertion losses in the transmitter and the receiver

D is the distance between the two adjacent APs

$u(d)$ and $v(d)$ are zero-mean Gaussian random variables, which model the shadow fading

The shadow fading is supposed to be auto-correlated with the auto-correlation function as follows:

$$F\{u(d_1)u(d_2)\} = \sigma_s^2 \exp(-|d_1 - d_2|/d_0) \tag{8.14}$$

where d_0 is the correlation distance and determines the fading rate of the correlation with the distance.

8.4.2 Optimal Handoff Location

As introduced in Section 8.3, when the train reaches the predetermined handoff location, the MIMO transmission mode will switch to multiuser multiplexing from the single-user STBC transmission mode, which will degrade the transmission performance. To determine the optimal handoff location for the multiple antenna configuration, we suppose that the handoff happens at the place where the FER of data transmission with the serving AP is same as the FER of transmission of the handoff signaling, which is transmitted between the train and the candidate AP. Because the FER is identical, the BER of the communication links is the same as well according to Equation 8.4. Therefore, we have

$$\alpha_A [\gamma_2(D - d)]^{-\beta_A} = \alpha_B [\gamma_1(d)]^{-\beta_B} \tag{8.15}$$

where:

$\gamma_2(D-d)$ is the receive SNR from AP_2 at the location where the distance is $(D-d)$ from AP_2

$\gamma_1(d)$ is the receive SNR from AP_1 with the distance of d

α_A, β_A, α_B, and β_B are defined as same as in Equation 8.3 with channel DoF A and B, respectively

The channel DoF applying zero forcing coding of MIMO multiplexing transmission is $2(N-M+1)$ for two antennas, where N is the number of receive antennas and M is the number of transmitting stations at the same time. And the DoF is $4(N-M+1)$ when applying STBC transmission for two antennas. When the number of antennas is four at the receiver, the DoF is $8(N-M+1)$ with STBC transmission, and it is $2(N-M+1)$ when multiplexing transmission is applied [27].

Suppose that the target packet error rate is 10^{-2}, we can calculate the target BER using Equation 8.4. In the handoff scenario, the channel DoF of the communication link between the train and the previous AP is 8, and it is 2 for the link between the train and the candidate AP where the handoff signaling and data packets are transmitted simultaneously. The required SNR from AP_2 is bigger than from AP_1 even with the same MCS level because the channel DoF of the link between AP_2 and the MS is different.

Given the two different required SNRs from the two APs, the received power from the two APs can be calculated by Equations 8.12 and 8.13. Then the distance from the optimal handoff location to the serving AP can be calculated from Equation 8.3:

$$d = D - D / \left[1 + 10^{\text{SNR}_{\text{diff}} / (10^* \text{PL})} \right] \tag{8.16}$$

where:

d is the distance from the optimal handoff location to the previous AP

D is the inter-site distance between the two adjacent APs

SNR_{diff} is the SNR difference between the required SNR with DoF = 8 and DoF = 2

PL is the path loss factor of the signal attenuation

8.4.3 Error-Free Period

Because handoff only happens at the predetermined position in the MAHO scheme, the number of handoffs is always 1. Therefore, the average error-free period can be obtained directly:

$$T_{\text{ef}} = D/v \tag{8.17}$$

where:

D is the inter-site distance between the successive APs

v is the average moving speed of the train

To analyze the error-free period of traditional handoff schemes, we suppose the triggering condition of existing schemes are as follows [30]:

1. The received signal strength at the candidate AP is greater than that of the serving AP by a hysteresis of h dB.
2. The average received signal strength at the serving AP is below an absolute threshold value T dBm.

The handoff will be triggered only when both the conditions are met. Because the handoff triggering positions are random due to the random receiving signal quality caused by the stochastic fading, it is difficult to calculate the instantaneous transmission interruption period in the handoff algorithms. To analyze and compare the handoff performance of existing "break-before-make" handoff scheme with our MAHO scheme, we will calculate the error-free period in average.

To get the error-free period of the communication system, we first calculate the number of handoffs when the MS moves from the serving AP to the next AP. The distance between the adjacent APs is divided into equal small intervals of length d_s. Suppose that the handoff will initiate at the end of each interval. Let $P_{ho}(k)$ be the handoff probability that when the MS is at the nth interval, P_{12} be the probability of the association changing from AP_1 to AP_2, P_{21} be the probability of the association changing from AP_2 to AP_1, $P_1(k)$ be the probability that the MS associates with AP_1, and $P_2(k)$ be the probability that the MS associates with AP_2. These probabilities are calculated recursively as follows:

$$P_{ho}(k) = P_1(k-1)P_{12} + P_2(k-1)P_{21} \tag{8.18}$$

$$P_1(k) = P_1(k-1)\left[1 - P_{12}(k)\right] + P_2(k-1)P_{21} \tag{8.19}$$

$$P_2(k) = P_2(k-1)\left[1 - P_{21}(k)\right] + P_1(k-1)P_{12} \tag{8.20}$$

where $k = 1,2,...,D/d_s$ and $P_1(0) = 1$ and $P_2(0) = 0$ are the initial values. The number of the handoffs from the serving AP to the next AP is the sum of the $P_{ho}(k)$ by k. In the calculation of $P_{ho}(k)$, if P_{12} and P_{21} are available, the results are obtained recursively. And then P_{12} and P_{21} are calculated as follows [30]:

$$P_{12}(k) \approx P\{k(d_k) < -h \mid x(d_{k-1}) \geq -h\} \times$$
$$P\{d(d_k < T) \mid x(d_k) < -h\} \tag{8.21}$$

$$P_{21}(k) \approx P\{x(d_k) > h \mid x(d_{k-1}) \leq h\} \times$$
$$P\{b(d_k < T) \mid x(d_k) > h\} \tag{8.22}$$

The number of handoffs N_{HO} is the sum of the handoff probability for all k's at the handoff points:

$$N_{HO} = \sum_k P_{ho}(k) \tag{8.23}$$

The handoff latency, at the level of 100 ms [25], is much less than the interval between the handoffs. When calculating the average error-free transmission period, it can be neglected. Therefore, the average error-free transmission period is

$$T_{ef} = \frac{D}{v \times N_{HO}} \tag{8.24}$$

where:
  v is the average moving speed of the train

8.4.4 FER of the Handoff Signaling

The SNR in dB at the receiver can be calculated as

$$SNR_{MAHO} = P_{MAHO} - P_N - NF \tag{8.25}$$

where:
  P_{MAHO} is the received power in dBm at the handoff location and can be calculated by Equation 8.12
  P_N is the power of the noise in dBm
  NF is the noise figure in dB at the receiver

The bit error rate of the handoff signaling can be calculated by Equation 8.3 with the channel DoF of 8, and then the FER of the handoff signaling can be obtained by Equation 8.4.

For traditional handoff schemes, the handoff may happen at random locations, so the FER of the handoff signaling cannot be obtained directly. To get the FER of the handoff signaling, same as in Section 8.4.3, we first get the probabilities $P_1(k)$ and $P_2(k)$ that the MS associates with AP_1 and AP_2, respectively, when the MS is at position k. The BER of the handoff signaling at position k can be calculated as follows:

$$P_{b\_ho}(k) = P_1(k) \times BER_{12} + P_2(k) \times BER_{21} \tag{8.26}$$

where:
  BER_{12} is the BER of handoff signaling which switches from AP_1 to AP_2 at position k
  BER_{21} is the BER from AP_2 to AP_1

The definitions of $P_1(k)$ and $P_2(k)$ are given in Equations 8.19 and 8.20. Then the average BER of the handoff signaling is

$$P_{\text{b\_ho}} = \frac{\sum_k \left[P_1(k) \times \text{BER}_{12} + P_2(k) \times \text{BER}_{21} \right]}{N_{\text{HO}}} \tag{8.27}$$

where N_{HO} is the number of handoff as defined in Equation 8.23. With the $P_{\text{b\_ho}}$, the FER of the handoff signaling can be calculated by using Equation 8.4.

8.4.5 Impacts on Ongoing Data Sessions

When the handoff happens, the MIMO transmission will switch to multiuser multiplexing mode from the STBC diversity mode, which will degrade the BER performance. As the data packets and handoff signaling are transmitted simultaneously by multiplexing, to guarantee the target FER, the receive SNR has to be increased, which will decrease the range of AP's coverage.

Another method to guarantee the FER performance is to adopt more robust MCS level when handoff happens. Because the handoff latency is in the level of 100 ms, the throughput loss in the handoff procedures due to the decrease of the data rate is neglectable. In this chapter, to simplify the analysis, we only consider the first case where no adaptive modulation and coding (AMC) schemes are adopted.

8.5 Simulation Results and Discussions

In this section, the simulation results on the handoff performance are given, which include the packer error rates of the handoff signaling, handoff latency, error-free period, and the FER of handoff signaling with different data rates.

8.5.1 Analysis of the Handoff Latency

In the MAHO scheme, the handoff procedure runs concurrently with data transmission and does not interrupt normal data transmission. The handoff delay can not describe the handoff performance exactly because the handoff procedure does not interrupt data transmission in our scheme. However, concurrent transmission of data packets with handoff signaling will increase the probability of the packet loss and then will increase the transmission delay of data packets due to the retransmission. Therefore, we have the delay increments of packet transmission when handoff (DIH) happens as the handoff performance measure. For "break-before-make" handoff schemes, the DIH value is the same as the handoff latency. The parameters used in the calculations are shown in Table 8.1.

Figure 8.5 shows the latency performance comparisons between the proposed MAHO scheme and the "break-before-make" handoff scheme. The handoff latency of traditional handoff schemes are calculated as in Equation 8.10, and the DIH is the difference of the handoff latency that the multiplexed transmission is used in

Table 8.1   Simulation Parameters

Notations	Value	Notations	Value
T_{DIFS}	38 μs [2]	Distance between two APs	600 m
T_{SIFS}	20 μs [2]	Transmit power at AP and MS	30 mW [25]
T_{ACK}	16 μs [2]	Insertion loss at AP	9 dB
aSlottime	9 μs [2]	Insertion loss at MS	7 dB
aCWmin	15 [2]	The height of the antenna at AP	4 m
T_{wait}	1000 ms [6]	The height of the antenna at MS	4 m
mChTime	10 ms	Standard deviation of shadow fading	5 dB
t_{Auth}	40 ms [16]	Correlation distance of shadow fading	25 m
Antenna gain at AP	13.5 dBi	Antenna gain at MS	9.5 dBi

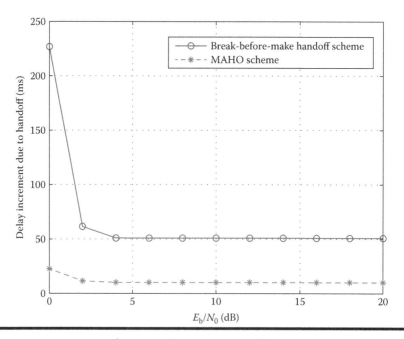

Figure 8.5   Latency performance improvements of the MAHO scheme.

MAHO scheme, with the handoff latency that multiplexing transmission is not implemented in traditional schemes. Although the DIH of traditional "break-before-make" handoff scheme is always above 50 ms, the DIH of the proposed MAHO scheme is almost negligible. This is because the FER is very small when the SNR at the receiver is greater than 4dB, even for MIMO multiplexing scheme. In CBTC communication system, the SNR at the receiver is always as high as more than 10 dB to guarantee the high reliability and robustness of the train control system. Therefore, the scheme can be reliably applied in CBTC communication system.

8.5.2 Error-Free Periods of Traditional Handoff Schemes

The time intervals of the different handoff schemes are given in Figure 8.6, and the cell layout parameters, which are from Beijing Subway Yizhuang Line, are shown in Table 8.1.

For the received signal strength-based handoff scheme, the mean number of handoffs is related to the hysteresis h and the receiving power threshold T. The number of handoffs decreases with the increase of the handoff hysteresis h. However, the big value of h will increase the delay time to initiate the handoff, which affects the transmission performance because the MS is communicating with the AP of worse signal quality. The number of handoffs increases as well when the handoff threshold gets larger from −70 to −60 dBm. When T is −70 dBm, the number of handoffs between two APs is near 1. The lower the handoff threshold T, the lesser the number of handoffs. This is because the lower signal strength threshold is easier to meet so that the handoff will not be triggered. However, the delay time to initiate handoff will also increase with the decrease of the handoff threshold, and this degrades the communication performance as well.

The number of handoffs in our proposed location-based MAHO scheme is the fixed 1 because the MS hands off to the candidate AP at the only fixed position between two APs. Therefore, the time interval between two consecutive handoffs is the time that the MS spends between the two handoff locations in the

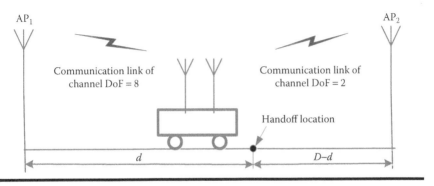

Figure 8.6 Time interval between two handoff procedures, $v = 80$ km/h.

travel, which is D/v. If the average speed of the train is 80 km/h, the error-free period in our scheme is about 27 s, and for the typical value of $h = 5$ dB, the period is about 20, 7, and 5.7 s with the handoff SNR thresholds of −70, −65, and −60 dBm, respectively. Our location-based MAHO scheme is superior to the existing schemes.

8.5.3 FER of Handoff Signaling with Different Data Rates

Although the data rate of the current CBTC system is less than 1 Mbps, future CBTC systems may require higher data rates. However, passenger information system (PIS) and video services of high data rates can be applied in the current urban railway transit communication system as well. Therefore, it is valuable to evaluate the handoff performance with higher data rates.

Figure 8.7 shows the FER comparisons of the handoff signaling between the MAHO and traditional handoff schemes with different data rates, where the handoff hysteresis is set as 5 dB and the absolute receive power threshold is set as 5 dB more than the minimum power requirements. In the simulation, we change the packet length to 500 bytes because the data rate increases to higher levels. When the data rate is 6.5 Mbps, the FER performance of handoff signaling is better than the traditional ones for almost all inter-site distances. We can see that the FER of MAHO signaling increases more rapidly than traditional schemes. This is because

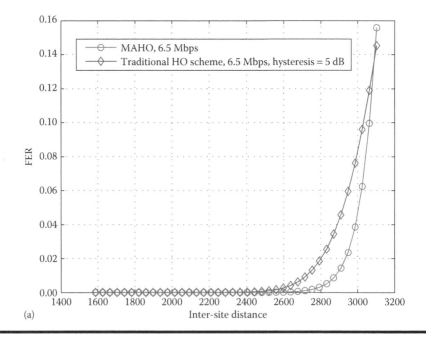

(a)

Figure 8.7 (a) FER at 6.5 Mbps. (*Continued*)

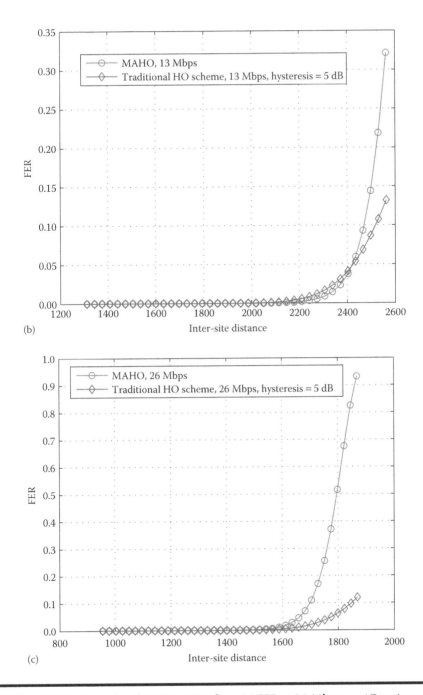

(b)

(c)

Figure 8.7 (Continued) **(b) FER at 13 Mbps. (c) FER at 26 Mbps.** *(Continued)*

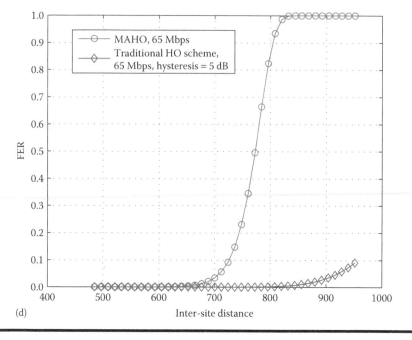

(d)   Inter-site distance

Figure 8.7 (Continued)   **(d) FER at 65 Mbps with different coverage areas.**

the FER of MAHO is affected by the FER of the data transmission multiplexed with handoff signaling, which is with lower DoF and then has more rapid increase of the error rate. When the data rate increases to 13 Mbps, the FER performance is much closer between the MAHO scheme and traditional HO scheme although the FER in MAHO is still a little better. When the data rate increases more, the FER performance of traditional HO schemes is better in the case of almost all the inter-site distances. This is because when the data rate increases, more increment of SNR in MAHO scheme is needed than traditional handoff schemes, and the FER of MAHO scheme will be larger than traditional schemes when the receive SNR is identical.

When the target FER of handoff signaling is 10^{-2}, without AMC applied, the maximum tolerable inter-site distances of different data rates are shown in Figure 8.8. We can see that when the data rate is the most possible 6.5 Mbps for CBTC service, the maximum inter-site distance between APs is 5% more than the traditional handoff schemes. It means that to meet the target FER, MAHO scheme needs less APs for CBTC service than traditional handoff schemes. However, when the data rate goes up to 13 Mbps, the difference of the inter-site distance will be closer. And when it is 26 Mbps or higher, the maximum tolerable inter-site distance in traditional handoff schemes is bigger than the MAHO.

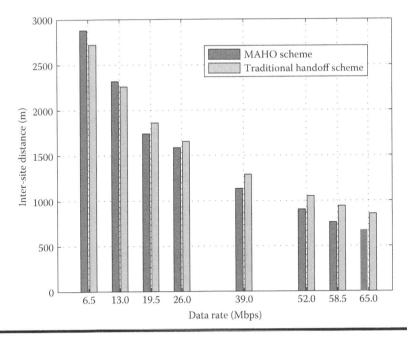

Figure 8.8 **Maximum inter-site distance to meet the HO FER.**

Therefore, if the required data rate is more than 13 Mbps, traditional handoff schemes require less APs to guarantee the FER of the handoff signaling.

8.6 Conclusion

Train–ground communication is one of the key technologies for the control center in the ground to control the train effectively in CBTC system. The WLAN handoff process has significant impacts on the train control performance in CBTC systems. In this chapter, we present a MIMO-assisted WLAN MAHO handoff scheme in CBTC communication system. Unlike existing handoff algorithms, we use location-based method to initiate the probe procedure. Handoff signaling and normal data packets are transmitted by different antennas, so that the handoff procedures can proceed without interrupting normal data transmissions. MIMO multiplexing mode and STBC interference cancelation algorithm are applied to the concurrent transmissions of train control information and handoff signaling. Simulation results show that the proposed MAHO scheme can reduce the handoff latency substantially, and the error-free period is also much larger compared with the existing schemes. When the data rate is no more than 13 Mbps, the FER of the handoff signaling is also lower than traditional handoff schemes.

References

1. H. Velayos and G. Karlsson. Techniques to reduce ieee 802.11b mac layer handover time. In *2004 IEEE International Conference on Communications*, Siena, Italia, June 20–24, 2004.
2. IEEE 802.11 Standards. Part 11: Wireless lan medium access control (mac) and physical layer (phy) specifications, 2012.
3. C.-S. Li, Y.-C. Tseng, and H.-C. Chao. A neighbor caching mechanism for handoff in ieee 802.11 wireless networks. In *International Conference on Multimedia and Ubiquitous Engineering*, Seoul, Korea, April 26–28, 2007.
4. I. Ramani and S. Savage. Syncscan: Practical fast handoff for 802.11 infrastructure networks. In *24th Annual Joint Conference of the IEEE Computer and Communications Societies. Proceedings IEEE*, Istanbul, Turkey, March 2005.
5. H. Jiang, B. Bing, and Z. Hongli. A novel handover scheme in wireless lan in cbtc system. In *IEEE International Conference on Service Operations, Logistics, and Informatics*, pages 473–477, Beijing, China, 2011.
6. L. Zhu, F. Richard Yu, B. Ning, and T. Tang. Cross-layer handoff design in mimo-enabled wlans for communication-based train control (cbtc) systems. *IEEE Journal on Selected Areas in Communications*, 30(4):719–728, 2012.
7. L. Zhu, F. Richard Yu, B. Ning, and T. Tang. Cross-layer design for video transmissions in metro passenger information systems. *IEEE Transactions on Vehicular Technology*, 60(3):1171–1181, 2011.
8. L. Zhu, F. Richard Yu, B. Ning, and T. Tang. Handoff performance improvements in mimo-enabled communication-based train control systems. *IEEE Transactions on Intelligent Transportation Systems*, 13(2):582–593, 2012.
9. V. Brik, A. Mishra, and S. Banerjee. Eliminating handoff latencies in 802.11 wlans using multiple radios: Applications, experience, and evaluation. In *Proceedings of the 5th ACM SIGCOMM Conference on Internet Measurement*, pages 3844–3848, Berkeley, CA, 2005.
10. K. Ramachandran, S. Rangarajan, and J.C. Lin. Make-before-break mac layer handoff in 802.11 wireless networks. In *2006 IEEE International Conference on Communications*, Istanbul, Turkey, June 2006.
11. Sunggeun Jin, Munhwan Choi, Ming-Deng Hsieh, and Sunghyun Choi. Multiple WNIC-based handoff in IEEE 802.11 WLANS. *IEEE Communications Letters*, 13(10):752–754, 2009.
12. Part 11: Wireless lan medium access control (mac) and physical layer (phy) specifications amennment 5: Enhancements for higher throughput, 2009.
13. H. Lee, H. Lee, and D.-H. Cho. Novel handover decision method in wireless communication systems with multiple antennas. In *Vehicular Technology Conference*, pages 1–5, Montréal, QC, Canada, 2006.
14. M. Cheng, X. Fang, and W. Luo. Beamforming and positioning-assisted handover scheme for long-term evolution system in high-speed railway. *Communications, IET*, 6(15):2335–2340, 2012.
15. P.W. Wolniansky, G.J. Foschini, G.D. Golden, and R.A. Valenzuela. V-blast: an architecture for realizing very high data rates over the rich-scattering wireless channel. In *URSI International Symposium on Signals, Systems, and Electronics, 1998.*, pages 295–300, Pisa, Italy, 1998.

16. B. Aboba. Fast handoff issues. Technical Report IEEE-03-155r0-I, IEEE 802.11 Working Group, March 2003.

17. EEIG ERTMS Users Group. Etcs/gsm-r quality of service ¨c operational analysis. Technical Report 04E117, ERTMS/ETCS, October 2005.

18. ALCATEL, ALSTOM, ANSALDO SIGNAL, BOMBARDIER, INVENSYS RAIL, and SIEMENS. Gsm-r interfaces class 1 requirements. Technical Report SUBSET-093 Version 2.3.0, ERTMS/ETCS, October 2005.

19. Standard for communications-based train control (cbtc) performance and functional requirements, 2004.

20. S.M. Alamouti. A simple transmit diversity technique for wireless communications. *IEEE Journal on Selected Areas in Communications*, 16(8):1451–1458, 1998.

21. A.F. Naguib. Combined interference suppression and frequency domain equalization for space-time block coded transmission. In *IEEE International Conference on Communications*, pages 3261–3266, Anchorage, AK, 2003.

22. A.F. Naguib, N. Seshadri, and A.R. Calderbank. Applications of space-time block codes and interference suppression for high capacity and high data rate wireless systems. In *Conference Record of the 32 Asilomar Conference on Signals, Systems & Computers*, pages 1803–1810, Pacific Grove, CA, 1998.

23. V. Tarock, N. Seshadri, and A.R. Calderbank. Space-time codes for high data rate wireless communications: performance criterion and code construction. *IEEE Transactions on Information Theory*, 44(2):744–765, 1998.

24. J. Kazemitaber and H. Jafarkhani. Multiuser interference cancellation and detection for users with more than two transmit antennas. *IEEE Transactions on Communications*, 56(4):574–583, 2008.

25. L. Zhang and H. Jiang. Technical report on the wireless free wave transmission in 52-53 sections in beijing subway. Technical report, State Key Laboratory of Rail Traffic Control and Safety of Beijing Jiaotong University, Beijing, China, August 2010.

26. T. Pollet, M. Van Bladel, and M. Moeneclaey. Ber sensitivity of ofdm systems to carrier frequency offset and wiener phase noise. *IEEE Transactions on Communications*, 43(2/3/4):191–193, 1995.

27. H. Jin, B.C. Jung, and D.K. Sung. A tradeoff between single-user and multi-user mimo schemes in multi-rate uplink wlans. *IEEE Transcations on Wireless Communications*, 10(10):3332–3342, 2011.

28. Richard Van Nee. Uplink mu-mimo sensitivity to power differences and synchronization errors. Technical Report IEEE-802.11-09/1036-00-00ac, IEEE 802.11 Working Group, September 2009.

29. D.B. Green and M.S. Obaidat. An accurate line of sight propagation performance model for ad-hoc 802.11 wireless lan (wlan) devices. In *Proc. IEEE ICC*, pages 3424–3428, New York, 2002.

30. N. Zhang and J.M. Holtzman. Analysis of handoff algorithms using both absolute and relative measurements. *IEEE Transactions on Vehicular Technology*, 45(1):174–179, 1996.

Chapter 9

Networked Control for a Group of Trains in Communications-Based Train Control Systems with Random Packet Drops

Bing Bu, F. Richard Yu, and Tao Tang

Contents

9.1 Introduction 178
9.2 Related Work 179
9.3 Trains' Control in CBTC Systems 180
    9.3.1 Communication Procedures with Packet Drops 180
    9.3.2 Trains' Control System and Equivalent NCS 181
        9.3.2.1 Trains' Control System in CBTC 181
        9.3.2.2 Equivalent NCS 183
    9.3.3 Analytical Formulation of CBTC 184
9.4 Packet Drops in Train–Ground Communications 186
    9.4.1 Packet Drops due to Random Transmission Errors 186
    9.4.2 Packet Drops due to Handover 187
        9.4.2.1 Handover Time 187

      9.4.2.2 AP's Coverage Area...188

      9.4.2.3 Overlapping Coverage Area...191

      9.4.2.4 Rate of Packet Drops Introduced by Handovers.............191

9.5   Trains' Control in CBTC with Packet Drops...193

      9.5.1 Currently Used Control Scheme in CBTC Systems.....................193

      9.5.2 States Estimation under Packet Drops...194

      9.5.3 Effects of Packet Drops on the Stability of the Trains'

           Control System 197

      9.5.4 Effects of Packet Drops on the Performances of the Trains'

           Control System 198

      9.5.5 Two Proposed Novel Control Schemes 199

9.6   Field Test and Simulation Results 201

      9.6.1 Field Test Results on the Packet Drop Rate 201

      9.6.2 Design of the Closed-Loop Control Systems.................................202

      9.6.3 Simulation Results of Trains' Control System Impacted by

           Packet Drops..203

9.7   Conclusion 208

References 210

9.1 Introduction

Rail-guided transport systems have attracted more and more attention, because they can provide greater transport capacity, superior energy efficiency, lower carbon emission, and outstanding features of punctuality and safety compared with other mass transit methods. The traditional rail system is a track-based train control system, which uses track circuit to coarsely determine the location of a train to transmit unidirectional ground–train control information. The coarse train positioning and low unidirectional communication throughput lead to low line capacity in TBTC rail systems. The typical minimum headway which is the time interval between two neighboring trains of TBTC is several minutes [1].

As a modern successor to TBTC, CBTC systems use continuous, high-capacity, bidirectional train–ground communication to transmit status and control commands of trains to realize automatic train control functions. The line capacity can be increased. The typical minimum headway of CBTC is 90 s or even less [1,2].

For urban transit systems, WLANs are commonly used due to the open standards and the available commercial off-the-shelf equipment [3]. Numerous WLAN-based CBTC systems have been deployed around the world, such as Beijing Metro Line 10 from Siemens [4] and Las Vegas Monorail from Alcatel [5]. However, WLANs are not originally designed for high-speed scenario; random transmission delays and packet drops are inevitable in train–ground communication.

Although the two key technologies of CBTC systems, trains' control and train–ground communication, are closely related, by now they are designed independently.

The requirement of trains' control on the quality-of-service (QoS) of train–ground communication and the impact of transmission errors on the performance of trains' control are still not clear. The separated design method limits the performance improvement of CBTC systems.

In this chapter, we integrate the design of trains' control and train–ground communication through modeling the control system of a group of trains in CBTC as a networked control system (NCS). We study the packet drops in CBTC system, introduce packet drops into the NCS model, analyze their impact on the stability and performance of CBTC systems, and propose two novel control schemes to improve the performances of CBTC system with random packet drops. The distinct features of this chapter are as follows:

1. We analyze packet drops in CBTC, caused by random transmission errors and handovers. The packet drop rate is formulated and related to handovers. The analytical results are well in line with the field test results.
2. Unlike the existing works that only consider a single train, we consider a group of trains in CBTC systems to improve the performances of CBTC.
3. We model the CBTC system with a group of trains as an NCS with random packet drops and analyze their impact on the stability and performances of CBTC.
4. We propose two control schemes for CBTC to minimize the energy consumption by reducing unnecessary traction and brake forces, to shorten the headway by decreasing distance fluctuation of trains under packet drops.
5. Extensive field test and simulation results are presented to show that our proposed schemes can provide less energy consumption, better riding comfortability, and higher line capacity compared to existing scheme.

The rest of this chapter is organized as follows: The related work is studied in Section 9.2. An introduction to CBTC systems is given in Section 9.3. The system to control a group of trains in CBTC is modeled and formulated as an NCS. Packet drops in CBTC systems are studied in Section 9.4. The effect of packet drops on the stability and the performances of trains' control are investigated and two control schemes to improve system performances are proposed in Section 9.5. Field test and simulation results are presented and discussed in Section 9.6. This chapter is concluded in Section 9.7.

9.2 Related Work

By now, the trains' control and train–ground communication in CBTC systems are designed independently. The control of a group of trains in CBTC does not consider too much of the train–ground communication issues. The traditional method is used [6–11]. A single train is usually considered based on the assumption that the distance between adjacent trains is long enough to ignore the interaction between them. However, this assumption does not hold in CBTC, which has a very tight headway. If a train's status cannot be sent to the train behind it in time due to transmission errors, the following train may need to trigger a brake or even emergency brake to ensure safety. A chain reaction might occur in the following trains, which seriously impairs the line capacity.

Moreover, traditional criteria are used in the design of train–ground communication system in CBTC, such as network capacity/throughput. Different schemes are proposed in [12–15] to improve the latency, availability, or throughput of train–ground communication. However, recent works in cross-layer design show that maximizing capacity/throughput does not necessarily benefit the upper layer which is train control in CBTC [16–21].

Our work is motivated by recent researches on NCS. NCS is a feedback control system wherein the control loops are closed through a real-time network [22]. The two main focuses in NCS-related researches are the effect of transmission delays and packet drops on the stability and the performance of control system [22–28]. In our chapter, the system to control a group of trains in CBTC is modeled as an NCS.

This work is also stimulated by researches on a platoon of vehicles' control in highway [29,30] because the control objectives of a string of vehicles, to minimize the space, velocity, and applied force errors [31], are similar to those of a group of trains in rail transit system.

9.3 Trains' Control in CBTC Systems

In this section, we first describe an overview of CBTC systems with packet drops. Then, we present the system to control a group of trains. At last, we model the trains' control system in CBTC as an NCS with packet drops in both uplink and downlink transmissions, followed by the analytical formulation of the CBTC system.

9.3.1 Communication Procedures with Packet Drops

A group of running trains on a line is depicted in Figure 9.1. The position and speed of the ith train at the beginning of the kth period are designated as s_k^i and v_k^i, respectively. The length of the ith train is defined as l^i. The distance between the tail of the ith train and that of the $(i-1)$th train is defined as d_k^i.

There is no direct or indirect data transmission among trains in CBTC systems. The following train can only get the status of the preceding train through LMAs received from ZC. Each train can get its own status precisely and timely through onboard odometer and speedometer. The communication procedures between ZC and trains with packet drops are illustrated in Figure 9.2. At the beginning of the $(k-1)$th period, the $(i-2)$th, $(i-1)$th, and ith trains send status to ZC separately.

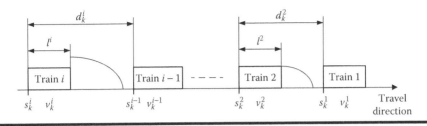

Figure 9.1 Train following model.

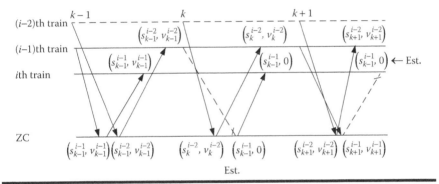

Figure 9.2 Communication procedure between ZC and the running trains with packet drops.

In the $(k-1)$th period, there is no packet loss. On receiving the status of the $(i-1)$th train, ZC sends s_{k-1}^{i-1} and v_{k-1}^{i-1} as LMA to the ith train and s_{k-1}^{i-2} and v_{k-1}^{i-2} are sent as LMA to the $(i-1)$th train.

Suppose that the packet from the $(i-1)$th train to ZC is dropped in the kth period. For safety, ZC assumes that the $(i-1)$th train keeps still since the last time ZC received its status. ZC uses the last available location of the $(i-1)$th train with a zero speed as LMA for the ith train.

In the $(k+1)$th period, the downlink packet from ZC to the ith train is lost. To ensure safe movement, the ith train assumes that the $(i-1)$th train maintains motionless since last time the ith train received LMA from ZC. The estimated LMA is used to calculate the speed/distance profile of the train.

9.3.2 Trains' Control System and Equivalent NCS

9.3.2.1 Trains' Control System in CBTC

The control system includes a group of trains, a ZC, and the network between ZC and trains, which is illustrated in Figure 9.3.

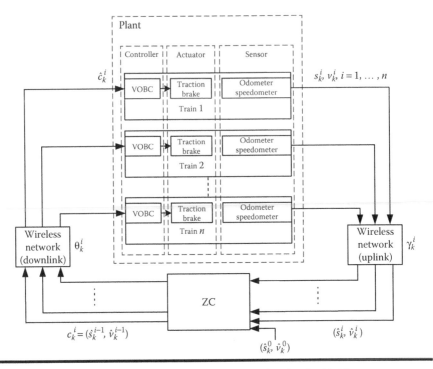

Figure 9.3   Model of system to control a group of trains in CBTC.

Throughout this chapter, unless otherwise stated, the superscript "i" denotes the ith train, and the subscript "k" indicates the kth period (time slot).

The result of uplink transmission of the status from the ith train to ZC is denoted as γ_k^i, where $\gamma_k^i = 0$ means that the uplink transmitted packet is dropped and $\gamma_k^i = 1$ indicates that the packet is successfully received by ZC. Due to packet drops in uplink transmissions, ZC can only get the estimated status of trains, $\hat{s}_k^i, \hat{v}_k^i$. The LMAs of the leading train are $\hat{s}_k^0$ and $\hat{v}_k^0$, and c_k^i is the LMA of the ith train. The result of downlink transmission from ZC to the ith train is denoted as θ_k^i. Due to packet drops in downlink transmissions, trains can only get the estimated LMA $\hat{c}_k^i$ to calculate the speed/distance profile.

There are two kinds of uncertainties in train–ground transmissions: transmission delay and packet drop. Due to the uncertainties in train–ground communications, retransmissions are needed. Random transmission errors lead to delays if packets are successfully received before reaching the retry limits. Otherwise, packets are dropped.

In WLANs, the station (STA) must break off the link with the original access point (AP) before associating with the objective AP, which introduces packet drops. The relationship between packet drops and handovers will be studied in Section 9.4.

9.3.2.2 Equivalent NCS

In Figure 9.4, we model the control system as an NCS with packet drops in both uplink and downlink transmissions. The group of trains composes the plant. The sensors are odometers and velocimeters of all the trains. All the vehicle onboard components (VOBCs) constitute the controller of the system. The actuator is all the traction and brake equipment of running trains.

Both uplink and downlink transmissions are diverse among trains due to various delays and packet drops. Some handover strategies are related to packet drops. A very large maximum number of retransmission times impair handover performance, because the STA may keep associating with the original AP even under a poor link quality. For such reason, a small maximum number of retry limits are usually adopted in CBTC systems, which means that delays introduced by random errors are very small. At the same time, trains and ZC are strictly time synchronized. On receiving the status of its preceding train, the train knows the exact sensor–controller delays, which can be estimated and compensated.

The main objectives of trains' control system are as follows:

■ To keep distances between trains to the scheduled headway to maximize line capacity
■ To control trains' velocities close to the maximum allowed speed to minimize the average travel time
■ To avoid unnecessary traction and brake to decrease energy consumption and make passengers comfortable

We define the system state as the distance and velocity deviations of all the trains, and define the output of the controller as the excess applied force. The status of trains is transmitted from odometers/speedometers to ZC and from ZC to VOBCs

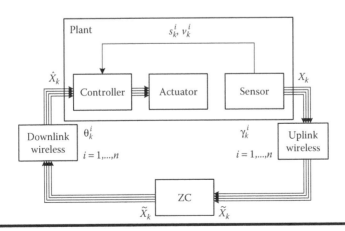

Figure 9.4 Equivalent networked control system.

because each train can directly get location and velocity of its own from onboard sensor. After receiving the status of the preceding train, the train generates control based on its distance and speed deviations. The uplink and downlink transmissions are equivalent to the transmission of system state.

As illustrated in Figure 9.4, X_k is the system state and γ_k^i is the result of uplink transmission of packet from the ith train to ZC. Due to packet drops in uplink transmission, ZC can only get the estimated state, $\tilde{X}_k$. ZC issues LMA, $\tilde{X}_k$. Because of packet drops in downlink transmission, indicated by θ_k^i, the controller can only get estimated LMA, $\hat{X}_k$.

9.3.3 Analytical Formulation of CBTC

To set up the analytical formulation of CBTC, we first consider the ideal case without packet drops. In CBTC systems, trains and ZC are strictly time synchronized and sampling with a very short period. We can consider that the system is discrete and linear time invariant.

$$X_{k+1} = AX_k + BU_k$$

$$U_k = -GX_k \tag{9.1}$$

where:

$X_k \in R^{2n \times 1}$ is the augmented system state
$U_k \in R^{n \times 1}$ is the output of the controller
$A \in R^{2n \times 2n}$ and $B \in R^{2n \times n}$ are real constant system matrices

The augmented system state includes distance and speed deviations of all the trains.

$$X_k = \begin{bmatrix} D_k \\ V_k \end{bmatrix}$$

$$D_k = \begin{bmatrix} \delta d_k^1 & \cdots & \delta d_k^n \end{bmatrix}$$

$$V_k = \begin{bmatrix} \delta v_k^1 & \cdots & \delta v_k^n \end{bmatrix} \tag{9.2}$$

$$U_k = \begin{bmatrix} \delta f_k^1 & \cdots & \delta f_k^n \end{bmatrix}$$

where:

D_k, V_k, and U_k are n-dimensional vectors
n is the number of trains
D_k is the deviation of distances between trains from the scheduled headways
V_k is the speed deviation from the optimal speed
U_k is the deviation of the applied forces from the basic resistance of trains

For every distance deviation in D_k, it has

$$d_k^i = s_k^{i-1} - s_k^i \tag{9.3}$$

$$\delta d_k^i = d_k^i - \Delta^i, i = 1,\ldots,n \tag{9.4}$$

where:
  d_k^i is the distance between the ith and $(i-1)$th trains
  δd_k^i is the distance deviation of the ith train

The optimal distance of the ith train is defined as Δ^i, which is a constant depending on the parameters of the ith train, such as emergency brake performance, length of the train, and safe margin.

The velocity deviation is defined as

$$\delta v_k^i = v_k^i - v \tag{9.5}$$

where:
  v_k^i is the train speed

We assume that the optimal speed v is a constant. To achieve the shortest travel time, v is just below the maximum speed with a reserved margin for safety.

For every deviation of applied force in U_k, it has

$$\delta f_k^i = f_k^i - g_k^i = m^i a_k^i \tag{9.6}$$

where:
  δf_k^i is the deviation of the applied force to the train from the basic resistance of the train
  f_k^i is the applied force to the ith train
  g_k^i is the basic resistance of the ith train

Based on kinematic equations, it has

$$\delta v_{k+1}^i = \delta v_k^i + a_k^i T \tag{9.7}$$

$$s_{k+1}^i = s_k^i + v_k^i T + \frac{1}{2} a_k^i T^2 \tag{9.8}$$

$$\delta d_{k+1}^i = \delta d_k^i + (\delta v_k^{i-1} - \delta v_k^i)T + \frac{1}{2}(a_k^{i-1} - a_k^i)T^2 \tag{9.9}$$

where:
  a_k^i is the acceleration of the ith train
  T is the sampling period of the sensor in the system

Based on Equations 9.7 through 9.9, we get matrices A and B in Equation 9.1.

9.4 Packet Drops in Train–Ground Communications

In this section, we first analyze packet drops introduced by random transmission errors. Then, we study the packet drops caused by handovers, which have a drop rate depending on the handover time, the AP coverage range, and the overlapping coverage area between APs.

9.4.1 Packet Drops due to Random Transmission Errors

There is an automatic repeat request (ARQ) scheme with carrier sense multiple access/collision avoidance (CSMA/CA) at the media access control (MAC) layer of IEEE 802.11. Usually, a small number of retransmission times are used in CBTC systems. The maximum number of retransmission times of Beijing Yizhuang and Changping Lines is less than 10.

We use MS Visual C++ 8.0 and IT++ 4.03 to build a link-level simulator of WLANs based on [32]. The frame error rate (FER) at different train speeds is given in Figure 9.5. The data rate is 6 Mbits/s and the packet size is 200 bytes. The simulated channel is flat fading channel with predefined train speeds. From Figure 9.5, we can see that the FER increases significantly with the growth of train speed.

Because of the random backoff scheme in CSMA/CA, the time interval between retransmissions (expect for the first two retransmissions) is much bigger than the channel

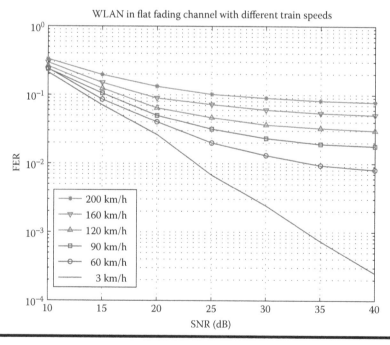

Figure 9.5 FER at certain train speeds.

coherent time. For ease of analysis, we assume that each retransmission is independent. Furthermore, we assume that the ACK transmission is lossless. For a given FER p and the maximum number of retransmission times r, the packet drop rate is

$$P = 1 - \sum_{j=0}^{r-1} p^j (1 - p) \qquad (9.10)$$

It can be seen from the FER in Figure 9.5 that the packet drop rate due to random transmission errors is very low for a fair signal-to-noise ratio. In Section 9.4.2, we will focus on the packet drops introduced by handover.

9.4.2 Packet Drops due to Handover

The onboard STA handovers at the edge of AP's coverage. In IEEE 802.11, handovers introduce transmission interruptions. The effects of handovers on packet drops depend on the handover time, the AP coverage area, and the overlapping coverage area between APs.

9.4.2.1 Handover Time

An active handover scheme is adopted in our CBTC system because it can realize shorter handover latency. In previous works [33–35], a handover process is divided into three phases: probing, authentication, and association. Probing delay is the dominant component. Since the data transmission is interrupted when the STA switches to other channels to scan APs and is resumed after successful negotiation of encryption keys with the target AP, the four-way handshake of encryption key should also be included in the analysis of transmission interruption time (handover time).

We develop a software to measure the handover time. A computer is connected to the backbone network, which has links to all the APs. Another computer is connected to the onboard STA. The ground computer sequentially and periodically sends data packets to the onboard computer. A unique sequence number is inserted into each packet. After handover, the onboard computer reorders the received packets and measures the handover time by counting the number of lost packets.

$$t_{ho} = n_d \cdot T_s \qquad (9.11)$$

where:
t_{ho} is the handover time
n_d is the number of dropped packets
T_s is the packet sending period

The field test results of handover time are shown in Figure 9.8, which are obtained from Beijing Yizhuang Lines.

9.4.2.2 AP's Coverage Area

The following factors should be considered to plan the AP's coverage area:

■ *Path loss.* In CBTC, the onboard and wayside antennas are installed only several meters high. The height of the receiving and transmitting antennas limits the transmission distance to a short range [36]. We use the empirical equation proposed by Green and Obaidat [37] to anticipate the free-space path loss. For the 2.4 GHz ISM band, it has

$$P_{ls} = 7.6 + 40\log_{10}d - 20\log_{10}h_t h_r \tag{9.12}$$

where:
P_{ls} is the path loss
d is the distance between the transmitter and receiver
h_t and h_r are the heights of transmitting and receiving antennas, respectively

■ *Shadow effect.* Both the local mean of envelope level Ω_r and the local mean of envelope power Ω_s obey log-normal distribution [38].

$$\Omega_r = <r(t)>, \Omega_s = <s(t)>, s(t) = r^2(t)/2$$

where:
$<\cdot>$ is the local mean function
$r(t)$ and $s(t)$ are the envelope and envelope power of received signal, respectively

The dB version of Ω_r and Ω_s obey Gaussian distribution with the same standard deviation σ_Ω.

The thresholds of the mean envelope power at certain possibilities are listed in Table 9.1.

■ *Multipath fading.* The envelope of the received signal obeys Rayleigh distribution. The cumulative distribution function of the envelope power of received signal can be expressed as a function of the fading factor [39].

$$F_s(x) = 1 - \exp\left(10^{V_F/10}\right) \tag{9.13}$$

where:
$V_F = 10\log_{10} x/\sigma^2$ is the fading factor
s is the envelope power of received signal
σ is the parameter of Rayleigh distribution

Table 9.1 Threshold of the Mean Envelope Power at Certain Possibilities ($\sigma_\Omega = 8$ dB)

x (in dB)	−20.6	−15.7	−13.2 α_8
$P(\Omega_p \leq x)$	0.005	0.025	0.05

Table 9.2 Threshold of the Envelope Power at Certain Probabilities

V_F (in dB)	−23	−16	−13 α_8
$P(s \leq V_F \cdot \sigma^2)$	0.005	0.025	0.05

When $V_F < 10$ dB, $F_s(x) \approx 10^{V_F/10}$

The thresholds of the envelope power at certain probabilities are given in Table 9.2.

■ *Receiver sensitivity.* The minimum rate-dependent input levels for receiver to meet required QoS performance is given in [32]. For 6 Mbits/s data rate, the minimum sensitivity is −82 dBm.

■ *Effective isotropic radiated power (EIRP).* In Europe and China, the maximum EIRP for 2.4 GHz ISM band is 100 mW ($G_t < 10$ dBi).

$$\text{EIRP} = P_t + G_t - P_{ti} \tag{9.14}$$

where:
P_t is the output power of transmitter
G_t is the antenna gain in dBi
P_{ti} is the insertion loss between transmitter and antenna

A fading margin is reserved in link budge for shadow fading and multipath fading to ensure the probability that the power of received signal at the edge of AP's coverage area exceeds the receiver's sensitivity and is above a given threshold. The received signal power is

$$P_r = \text{EIRP} - P_{ls} - P_{sf} - P_{mf} + G_r - P_{ri} \tag{9.15}$$

where:
P_r is the received power at the cell's edge
P_{sf} is the power loss induced by shadow fading
P_{mf} is the power loss due to multipath fading
G_r is the gain (in dBi) of receiving antenna
P_{ri} is the insertion loss at the receiver

The selected parameters for AP's coverage plan are listed in Table 9.3. The curves of different probabilities that the received power at a given distance exceeds a certain level are depicted in Figure 9.6. As shown in the figure, when the cell coverage range is less than 200 m, the probability that the power of received signal would exceed −82 dBm is above 95%.

Table 9.3 Parameters for AP Coverage Plan

Parameters	Value α_g
EIPR (dBm)	17
h_t (m)	4.5
h_r (m)	5
G_r (dBi)	10
P_{ri} (dB)	4

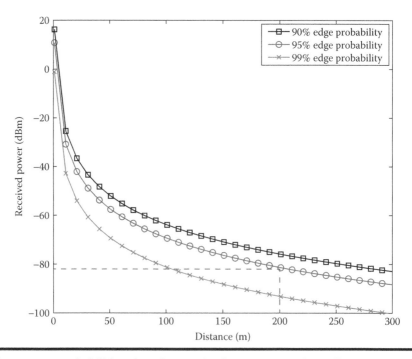

Figure 9.6 Probabilities that the received power at a given distance exceeds certain levels.

9.4.2.3 Overlapping Coverage Area

We propose a method to determine the overlapping coverage area of APs in CBTC. As illustrated in Figure 9.7, APs are linearly deployed along the track. Two directional antennas are installed and connected to each wayside AP and to each onboard STA. At first, the STA associates with AP1. At point "a," the STA detects the signal of AP2. At point "b," the strength of signal from AP2 exceeds that from AP1. The STA initiates a handover. Normally, the STA sends de-association frame to the original AP to release resources after successfully negotiating encryption keys with the objective AP. The STA should complete the whole handover process before running out of the coverage area of AP1. The length of overlapping coverage area must satisfy the following:

$$d_{ov} = d_{ab} + d_{bc}$$

(9.16)

$$d_{ab} = d_{bc} \geq v_m \cdot t_{mh}$$

where:
    d_{ov} is the length of the overlapping coverage area
    d_{ab} is the distance between point "a" and "b"
    d_{bc} is the distance between point "b" and "c"
    v_m is the maximum train speed
    t_{mh} is the longest handover time

The field test results on handover time are given in Figure 9.8. The rarely appeared (<1%) overlong t_{ho} are removed as bad values. It is assumed that $t_{mh} = 180$ ms. In the design of AP's overlapping coverage area, a maximum train speed of 200 km/h is considered for CBTC's future applications, and $d_{ov} = 20$ m.

9.4.2.4 Rate of Packet Drops Introduced by Handovers

Based on AP's coverage range, overlapping coverage area, and the handover time, we can get the rate of packet drops introduced by handover. Suppose that there

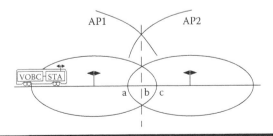

Figure 9.7 Overlapping coverage area of APs.

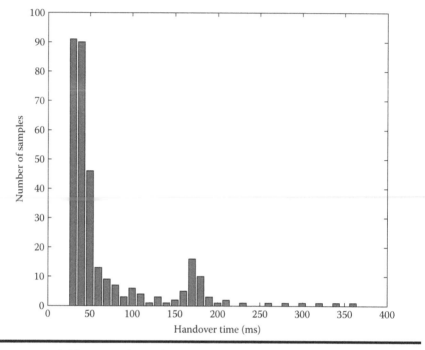

Figure 9.8   Field test results on handover time.

are n APs with $n-1$ overlapping coverage areas. The running time of a train at speed v_m is

$$t_r = \frac{d_{\text{ap}} \cdot n - d_{\text{ov}} \cdot (n-1)}{v_m} \tag{9.17}$$

where:

d_{ap} is the coverage range of an AP

To make the problem tractable, we make the following simplifications:

- For business operation speed, handovers introduce the average handover time t_{ah}.
- There is no ping-pong handover. STA makes only one handover in the overlapping coverage area.
- The number of packet drops caused by each handover is

$$n_{\text{ho}} = \left\lceil \frac{t_{\text{ah}}}{T} \right\rceil$$

where:

t_{ah} is the average handover time

$\left\lceil \dfrac{t_{\text{ah}}}{T} \right\rceil$ is the minimum integer that is greater than or equal to $\left\lceil \dfrac{t_{\text{ah}}}{T} \right\rceil$

Table 9.4   Parameters to Analyze the Rate of Packet Drops Introduced by Handovers

Parameters	Value α_g
d_{ap} (m)	200
d_{ov} (m)	20
t_{mh} (ms)	180
t_{ah} (ms)	67
T (ms)	300
v_m (km/h)	200

The number of total transmitted packets is $n_{to} = \left\lceil \dfrac{t_r}{T} \right\rceil$. The rate of packet drops introduced by handovers is

$$P_{ho} = \lim_{n \to \infty} \frac{(n-1)n_{ho}}{n_{to}} \tag{9.18}$$

Based on the field test results on handover time, $t_{ah} = 67$ ms. Taking parameters from Table 9.4, we can obtain the rate of packet drops introduced by handovers at a train speed of 60 km/h, $P_{ho}(60) \approx 0.027$. For future CBTC applications with a speed of 200 km/h, we assume that handovers always introduce the maximum communication interruption time, $t_{mh} = 180$ ms. It has $P_{ho}(200) \approx 0.1$.

9.5 Trains' Control in CBTC with Packet Drops

In this section, packet drops are first introduced into the trains' control system in CBTC. Then, the effects of packet drops on state transmission and estimation are studied. Their impacts on stability and performances of trains' control system are also analyzed. Finally, we propose two novel control schemes to improve the performances of trains' control system under packet drops.

9.5.1 Currently Used Control Scheme in CBTC Systems

In current CBTC systems, a train uses the status of its own and the previous train to generate control commands. Each train can directly obtain its status through onboard sensors without impairment from the train–ground communication.

If the packet containing the status of the preceding train is lost, the following train uses the real status of its own and the estimated status of the previous train to generate control commands. For safety, the following train assumes that the preceding train keeps still at the last received position.

9.5.2 States Estimation under Packet Drops

The handover of onboard STA may affect either or both the uplink and downlink transmissions. We assume that γ_k^i and θ_k^j obey independent and identical Bernoulli distribution.

$$P\left(\gamma_k^i = 1\right) = P\left(\theta_k^j = 1\right) = p$$

$$P\left(\gamma_k^i = 0\right) = P\left(\theta_k^j = 0\right) = 1 - p$$

The jth train's estimated status of the ith train depends on the results of the uplink transmission from the ith train to ZC and the downlink transmission from ZC to the jth train.

$$\hat{s}_k^{i,j} = \vartheta_k^{i,j} s_k^i + \overline{\vartheta}_k^{i,j} \hat{s}_{k-1}^{i,j} \tag{9.19}$$

$$\hat{v}_k^{i,j} = \vartheta_k^{i,j} v_k^i + \overline{\vartheta}_k^{i,j} (\hat{v}_{k-1}^{i,j} + \Delta_v) \tag{9.20}$$

where:

$\hat{s}_k^{i,j}$ and $\hat{v}_i^{i,j}$ are the jth train estimated location and the speed of the ith train, respectively

Δ_v is the maximum increase of speed a train may experience in one period τ

$\vartheta_k^{i,j}$ indicates that if the jth train received the status of the ith train

Here, we only consider the case that the following train uses the status of the foregoing trains, $i < j$.

$$\vartheta_k^{i,j} = \begin{cases} \gamma_k^i \theta_k^j, i < j \\ \vartheta_k^{i,i} = 1, i \geq j \end{cases} \tag{9.21}$$

$$\overline{\vartheta}_k^{i,j} = 1 - \vartheta_k^{i,j}$$

For the currently adopted control scheme in CBTC systems, the status of the ith train is only sent to the $(i+1)$th train to generate control, $j = i + 1$.

We define the jth train estimated system state as

$$\hat{X}_k^j = \begin{bmatrix} \hat{D}_k^j & \hat{V}_k^{j'} \end{bmatrix} \tag{9.22}$$

where:

$\hat{D}_k^j$ and $\hat{V}_k^j$ are n-dimensional vectors of deviations of the estimated distances and velocities

$$\hat{D}_k^j = \begin{bmatrix} \delta \hat{d}_k^{1,j} & \cdots & \delta \hat{d}_k^{i,j} & \cdots & \delta \hat{d}_k^{n,j} \end{bmatrix}$$

$$\hat{V}_k^j = \begin{bmatrix} \delta \hat{v}_k^{1,j} & \cdots & \delta \hat{v}_k^{i,j} & \cdots & \delta \hat{v}_k^{n,j} \end{bmatrix}$$

where:

$\delta \hat{d}_k^{i,j}$ and $\delta \hat{v}_k^{i,j}$ are deviations of the jth train estimated distance and velocity of the ith train, respectively

If the status of the ith train is successfully received by the jth train, it will be used to get $\delta \hat{d}_k^{i,j}$ and $\delta \hat{v}_k^{i,j}$. Otherwise, the jth train uses the last estimated distance and speed of the ith train, and considers the worst case to update the deviation of the estimated distance and velocity. $\delta \hat{d}_k^{i,j}$ and $\delta \hat{v}_k^{i,j}$ depend on γ_k^{i-1}, γ_k^i, and θ_k^j, and can be expressed as

$$
\begin{aligned}
\delta \hat{d}_k^{i,j} &= \vartheta_k^{i,j}(\hat{s}_k^{i-1,j} - s_k^i - \Delta^i) + \bar{\vartheta}_k^{i,j}(\delta \hat{d}_{k-1}^{i,j} - \Delta_d) \\
&= \vartheta_k^{i,j}\left(\vartheta_k^{i-1,j}\delta d_k^i - \frac{\bar{\vartheta}_k^{i-1,j}T}{2}\delta v_k^i \right) \\
&\quad - \vartheta_k^{i,j}\frac{\bar{\vartheta}_k^{i-1,j}T}{2}(\delta v_{k-1}^i + 2v) \\
&\quad + (1 - \vartheta_k^{i,j}\vartheta_k^{i-1,j})\delta \hat{d}_{k-1}^{i,j} - \vartheta_k^{i,j}\Delta_d
\end{aligned}
\tag{9.23}
$$

$$
\delta \hat{v}_k^{i,j} = \vartheta_k^{i,j}\delta v_k^i + \bar{\vartheta}_k^{i,j}(\delta \hat{v}_{k-1}^{i,j} + \Delta_v)
\tag{9.24}
$$

where:

Δ_d is the maximum distance a train may travel during one period T

Δ_v is the maximum increase of speed a train may experience during one time slot

We omit the item $\vartheta_k^{i,j}\bar{\vartheta}_k^{i-1,j}vT$, because vT is very small compared with Δ^i. It can be included in Δ^i for safety. For algebraic operation, we define the following diagonal matrix to indicate the uplink and downlink transmissions:

$$\Theta_k^j = \text{Diag}\left(\begin{bmatrix} \vartheta_k^{1,j} & \cdots & \vartheta_k^{n,j} \end{bmatrix} \right)$$

$$\bar{\Theta}_k^j = \text{Diag}\left(\begin{bmatrix} \bar{\vartheta}_k^{1,j} & \cdots & \bar{\vartheta}_k^{n,j} \end{bmatrix} \right)$$

$$\Upsilon_k^j = \text{Diag}\left(\begin{bmatrix} \vartheta_k^{0,j} & \cdots & \vartheta_k^{n-1,j} \end{bmatrix} \right)$$

$$\overline{Y}_k^j = \mathrm{Diag}\left(\left[\begin{array}{ccc} \overline{9}_k^{0,j} & \cdots & \overline{9}_k^{n-1,j} \end{array}\right]\right)$$

where:

Θ_k^j, $\overline{\Theta}_k^j$, Y_k^j, and $\overline{Y}_k^j \in R^{n \times n}$

Diag(X) generates a matrix, the main diagonal of which are components of vector X

From Equations 9.23 and 9.24, the *j*th train estimated system state can be expressed as

$$\hat{X}_k^j = \hat{C}_k^j X_k + \hat{E}_k^j X_{k-1} + \hat{F}_k^j \hat{X}_{k-1}^j + \hat{H}_k^j W_0^j$$

$$\hat{C}_k^j = \begin{bmatrix} \Theta_k^j Y_k^j & -\dfrac{T}{2}\Theta_k^j \overline{Y}_k^j \\ 0_{n,n} & \Theta_k^j \end{bmatrix}, \hat{E}_k^j = \begin{bmatrix} 0_{n,n} & -\dfrac{T}{2}\Theta_k^j \overline{Y}_k^j \\ 0_{n,n} & 0_{n,n} \end{bmatrix}$$

$$\hat{F}_k^j = \begin{bmatrix} I_n - \Theta_k^j Y_k^j & 0_{n,n} \\ 0_{n,n} & \Theta_k^j \end{bmatrix}, \hat{H}_k^j = \begin{bmatrix} \overline{\Theta}_k^j & 0_{n,n} \\ 0_{n,n} & \overline{\Theta}_k^j \end{bmatrix}$$

(9.25)

$$W_0^j = \begin{bmatrix} -\Delta_d & \cdots & -\Delta_d & \Delta_v & \cdots & \Delta_v \end{bmatrix}'$$

For a group of trains' analysis in CBTC systems, we define the following matrices:

$$\hat{C}_k = \mathrm{BlkDiag}\left(\left[\begin{array}{ccc} \hat{C}_k^1 & \cdots & \hat{C}_k^n \end{array}\right]\right)$$

$$\hat{E}_k = \mathrm{BlkDiag}\left(\left[\begin{array}{ccc} \hat{E}_k^1 & \cdots & \hat{E}_k^n \end{array}\right]\right)$$

$$\hat{F}_k = \mathrm{BlkDiag}\left(\left[\begin{array}{ccc} \hat{F}_k^1 & \cdots & \hat{F}_k^n \end{array}\right]\right)$$

$$\hat{H}_k = \mathrm{BlkDiag}\left(\left[\begin{array}{ccc} \hat{H}_k^1 & \cdots & \hat{H}_k^n \end{array}\right]\right)$$

$$W_0 = \left(\left[\begin{array}{ccc} W_0^{1'} & \cdots & W_0^{n'} \end{array}\right]\right)$$

where:

BlkDiag([···]) constructs block diagonal matrix from input arguments

The system state, estimated system state, parameter matrix, closed-loop gain, and control vector are defined as follows:

$$Y_k = \begin{bmatrix} X'_k & \cdots & X'_k \end{bmatrix}', A_y = \mathrm{BlkDiag}\big(\begin{bmatrix} A & \cdots & A \end{bmatrix}\big)$$

$$\hat{Y}_k = \begin{bmatrix} \hat{X}^{1'}_k & \cdots & \hat{X}^{n'}_k \end{bmatrix}', B_y = \mathrm{BlkDiag}\big(\begin{bmatrix} B & \cdots & B \end{bmatrix}\big)$$

$$\hat{G}_k = \mathrm{BlkDiag}\big(\begin{bmatrix} G_k & \cdots & G_k \end{bmatrix}\big), \hat{U}_k = \begin{bmatrix} U'_k & \cdots & U'_k \end{bmatrix}'$$

The system model of a group of trains' control with variant uplink and downlink packet drops can be expressed as

$$
\begin{aligned}
Y_{k+1} &= A_y Y_k - B_y \hat{G}_k \hat{Y}_k \\
&= A_y Y_k - B_y \hat{G}_k (\hat{C}_k Y_k + \hat{E}_k Y_{k-1} + \hat{F}_k \hat{Y}_{k-1} + \hat{H}_k W_0) \quad (9.26) \\
\hat{U}_k &= -\hat{G}_k \hat{Y}_k
\end{aligned}
$$

9.5.3 Effects of Packet Drops on the Stability of the Trains' Control System

To analyze the effects of packet drops on the stability of the trains' control system, an augmented state is defined as

$$Z_k = \begin{bmatrix} Y'_k & Y'_{k-1} & \hat{Y}'_{k-1} & W'_0 \end{bmatrix} \quad (9.27)$$

The augmented closed-loop system is

$$Z_{k+1} = \Phi_k Z_k$$

$$
\Phi_k = \begin{bmatrix}
A_y - B_y G_k \hat{C}_k & -B_y G_k \hat{E}_k & -B_y G_k \hat{F}_k & -B_y G_k \hat{H}_k \\
I_{2n} & 0_{2n,2n} & 0_{2n,2n} & 0_{2n,2n} \\
\hat{C}_k & \hat{E}_k & \hat{F}_k & \hat{H}_k \\
0_{2n,2n} & 0_{2n,2n} & 0_{2n,2n} & I_{2n}
\end{bmatrix} \quad (9.28)
$$

where:
Φ_k is a time variable matrix depending on Υ_k^j and Θ_k^j

If we assume the uplink and downlink transmissions are independent, $\Theta_k^j \Upsilon_k^j$ has 2^n alternatives. Accordingly, there are $2^{2n-1}+1$ options for Φ_k, $\Phi_k \in \{\Phi_s, s = 1,\ldots,2^{2n-1}+1\}$. The occurrence rate of Φ_s is designated as r_s; thus, $\sum_{s=1}^{2^{2n-1}+1} r_s = 1$.

The system of Equation 9.28 falls under the class of asynchronous dynamical systems (ADSs) [23]. The stability of ADS is given by the following theorem:

Theorem 1: *For the asynchronous dynamical system*

$$Z_{k+1} = \Phi_k Z_k$$

where:

$\Phi_k \in \{\Phi_s, s = 1,\cdots,N\}$ *is the coefficient matrix at the kth period.* Φ_s *occurs with rate* r_s, $\sum_{s=1}^N r_s = 1$.

The system is exponentially stable with decay rate of $\alpha^{0.05}$, *if there exists a Lyapunov function* $V(x)$, $R^n \to R_+$:

$$\beta_1 \parallel x \parallel^2 \le V(x) \le \beta_2 \parallel x \parallel^2$$

where $\beta_1,\beta_2 > 0$ *and* V *satisfies the following conditions:*
■ *There exists* $\alpha_s > 0, s = 1,\cdots,N$, *such that*

$$V(x_{k+1}) - V(x_k) \le (\alpha_s^{-0.1} - 1)V(x_k)$$

■ *The* α_s *satisfies* $\prod_{s=1}^N \alpha_s^{r_s} > \alpha > 1$.

The method to prove Theorem 1 is similar to that in [22]. We use a smaller exponent of α_s in the definition of the Lyapunov function, because it makes the linear matrix inequality (LMIs) more feasible under high packet drop rate.

9.5.4 Effects of Packet Drops on the Performances of the Trains' Control System

It is assumed that deviations of distance, velocity, and applied force without transmission errors are the optimal values. The total cost, cost of divagation of distance, velocity, and applied force from the optimal values are defined as follows:

$$J_N = \alpha_d J_{N,\delta d} + \alpha_v J_{N,\delta v} + \alpha_f J_{N,\delta f} \tag{9.29}$$

$$J_{N,\delta d} = \sum_{k=1}^{N} \sum_{i=1}^{n} (\delta d_k^i - \delta \tilde{d}_k^i)^2$$

$$J_{N,\delta v} = \sum_{k=1}^{N} \sum_{i=1}^{n} (\delta v_k^i - \delta \tilde{v}_k^i)^2$$

$$J_{N,\delta f} = \sum_{k=1}^{N} \sum_{i=1}^{n} (\delta f_k^i - \delta \tilde{f}_k^i)^2$$

where the optimal distance, velocity, and applied force deviations are designated as $\delta \tilde{d}_k^i$, $\delta \tilde{v}_k^i$, and $\delta \tilde{f}_k^i$, respectively. The total cost is defined as J_N, $J_{N,\delta d}$ is the cost of divagation of distance deviation from $\delta \tilde{d}_k^i$, $J_{N,\delta v}$ is the cost of divagation of velocity deviation from $\delta \tilde{v}_k^i$, and $J_{N,\delta f}$ is the cost of divagation of applied force deviation from $\delta \tilde{f}_k^i$. The weight coefficients are defined as α_d, α_v, and α_f, respectively.

The power of a train is proportional to the applied force at any given speed. The applied forces without packet drops are the most energy-saving ones, because they only overcome the basic resistance of trains in steady state. To reduce energy consumption of trains' control in CBTC with packet drops is to decrease unnecessary traction and brake and keep the applied forces of trains close to the optimal curves. Therefore, a smaller $J_{N,\delta f}$ means less energy consumption. The line capacity is inversely proportional to the headway. As $J_{N,\delta d}$ indicates the fluctuation of train's distance from the one in steady state, a shorter headway is needed to eliminate the interaction between adjacent trains for a smaller $J_{N,\delta d}$, which implies a higher line capacity.

9.5.5 Two Proposed Novel Control Schemes

Previously, we modeled the control of a group of trains in CBTC system as an NCS with packet drops. The packet drop rate in CBTC systems is formulated and related to handovers. The impact of packet drops on the estimated system state is analyzed. After that, packet drops are introduced into the NCS model and cost functions are defined to analyze the impact of packet drops on the stability and performances of the trains' control system in CBTC. Our previous work has clarified the two unclear questions we mentioned in Section 9.1. The QoS of train–ground communication should satisfy the stability and performance requirement of trains' control system. Also, the effect of packet drops on the trains' control system can be analyzed through our method. In the next, we will use the NCS model to design the control scheme. By doing so, the design of trains' control is integrated with the train–ground communication to improve the performances of trains' control under packet drops in CBTC systems.

For the current control scheme, only the status of the previous train is used for the following train to create control command. In this section, we propose two control schemes to improve the performances of trains' control system under packet drops. As ZC has status of all the running trains, we propose to use all or some of the forgoing trains' status for the following train to generate control command. Therefore, it can respond more rapidly to the status change of foreceing trains. Due to packet drops, the following train may receive the real or estimated status of the forgoing trains; it can select to use all or some of the status to keep its performances close to the optimal values. The proposed schemes will be verified in Section 9.6 by simulation results. The proposed schemes are as follows:

- From the beginning of each period, ZC waits to receive the status of all the trains before sending LMAs. ZC's waiting time is less than a predefined uplink reception interval. The equivalent uplink transmission delay is uniform among trains.
- If a train's status is successfully received by ZC, it will be inserted into the LMA for the following trains. Otherwise, the estimated status of the train will be used.
- ZC sends the status or estimated status of all the foregoing trains together in one packet to a specific train to generate control commands.
- The controller waits for a predefined duration from the beginning of each period to generate and output control commands to the actuators. The length of the duration τ is no less than $2\tau_m$. Here, τ_m is the delay introduced by the maximum number of retransmissions.
- If the LMA containing the status of all the foregoing trains is available when the controller generates and sends controls to the actuator, it will be used to calculate the control commands. Otherwise, an estimated LMA will be used instead.
- Based on the status or estimated status of the foregoing trains, each train selects the closed-loop gain to minimize the fluctuation of applied force or distance deviations around the optimal values. Each train selects its closed-loop gain by the following two criteria:
  - *Minimized force criterion.* To minimize its cost of divagation of the applied force deviations from the optimal values. G_s is the combination of the closed-loop gains of all the trains.

$$\min \sum_{i=1}^{n} (\delta f_k^i - \delta \tilde{f}_k^i)^2 \tag{9.30}$$

  - *Minimized distance criterion.* To minimize its cost of divagation of the distance deviations from the optimal values.

$$\min \sum_{i=1}^{n} (\delta d_k^i - \delta \tilde{d}_k^i)^2 \tag{9.31}$$

G_k of $\hat{G}_k$ in (9.26) is time variable. It is a combination of closed-loop gains of all the trains. G_k has multiple options, $G_k \in \{G_s, s = 1, ..., M\}$.

9.6 Field Test and Simulation Results

In this section, we first give field test results on the rate of packet drops introduced by handovers in a real CBTC system. Then, we design a control system of three trains, and the two novel control schemes are evaluated. Finally, simulation results are presented, and the effects of packet drops on the stability and performances of trains' control system are discussed.

9.6.1 Field Test Results on the Packet Drop Rate

We measure the packet drop rate on Beijing Yizhuang Subway Line. Yizhuang Line is a typical CBTC line that began its commercial operation at the end of 2010. It is 23 km in length. One hundred and eight APs are installed in the main line.

We use self-developed software, mentioned in Section 9.4, to obtain the packet drop rate introduced by handover. Field test results at 60 km/h speed are listed in Table 9.5.

In order to get enough samples, we transmit packets with a much shorter period, $T = 16$ ms. The packet drop rate is between 0.003 and 0.005, which is close to the analytical result we can obtain using the method given in Section 9.4, $T = 16$ ms, $P_{ho} = 0.0074$. The field test results prove the validity of our proposed analytical model of packet drop rate in CBTC. Then, we use the analytical model to get the packet drop rate at a very low speed (30 km/h), $P_{ho}(30) \approx 0.01$, and at a much higher speed (200 km/h), $P_{ho}(200) \approx 0.1$. According to the two drop rates, random packet drops will be generated to analyze the performances of different control schemes in the simulations described in Sections 9.6.2 and 9.6.3.

Table 9.5 Field Test Results on the Rate of Packet Drops Introduced by Handovers

Test Duration (s)	Dropped Packets	Total Packets	Drop Rate α_g
2448	517	145975	0.003542
2378	466	133615	0.003488
2748	510	162783	0.003133
4749	1098	222766	0.004929
1719	399	102208	0.003904

9.6.2 Design of the Closed-Loop Control Systems

A control system of three trains is designed. Train 1 (T1) is the leading train. Train 3 (T3) is the last train. Train 2 (T2) is in the middle. The main parameters used in our simulations are listed in Table 9.6.

We design the closed-loop control systems based on the following control schemes.

"Short view" scheme (Sv). The following train uses the speed and velocity information of its own and its previous train to generate control commands.

"Long view and minimized distance" scheme (Lv_d). Each train uses different combinations of the status or estimated status of foregoing trains to generate control commands by selecting its closed-loop gain to minimize the divagation of its distance deviations from the optimal values.

"Long view and minimized force" scheme (Lv_f). Each train selects its closed-loop gain to minimize the divagation of its applied force deviations from the optimal values.

The Sv scheme is based on the currently adopted scheme in CBTC systems. The Lv_d and Lv_f schemes are based on our proposed methods.

The system of Equation 9.28 is controllable and observable. The poles placement method is used to design the closed-loop gains [40,41]. The leading train has only one closed-loop gain.

$$G^{1,1} = \begin{bmatrix} -0.0820 & 0 & 0 & 0.6845 & 0 & 0 \end{bmatrix}$$

Train 2 has two alternatives.

$$G^{2,1} = \begin{bmatrix} 0 & -1.2026 & 0 & 0 & 0.9016 & 0 \end{bmatrix}$$

$$G^{2,2} = \begin{bmatrix} -0.0449 & -0.4326 & 0 & -0.07157 & 0.7707 & 0 \end{bmatrix}$$

Table 9.6 Simulation Parameters

Parameters	Value α_g
Length of train (m)	118
Mass of train	1
Optimal speed (km/h)	90
Maximum normal brake rate (m/s²)	1.5
Emergency brake rate (m/s²)	2
Maximum acceleration rate (m/s²)	2
Maximum input distance error (m)	15

Train 3 has four options.

$$G^{3,1} = \begin{bmatrix} 0 & 0 & -1.5023 & 0 & 0 & 0.8032 \end{bmatrix}$$

$$G^{3,2} = \begin{bmatrix} -0.0382 & -0.0380 & -0.3850 & \cdots \end{bmatrix}$$

$$-0.07331 \quad -0.0783 \quad 0.7048 \end{bmatrix}$$

$$G^{3,3} = \begin{bmatrix} -0.0460 & 0 & -0.4629 & -0.07333 & 0 & 0.7048 \end{bmatrix}$$

$$G^{3,4} = \begin{bmatrix} 0 & -0.1035 & -1.0360 & 0 & -0.07333 & 0.7048 \end{bmatrix}$$

G_s are combinations of closed-loop gains of all the trains. For the control system of three trains, there are eight selections.

$$G_1 = \begin{bmatrix} G^{1,1'} & G^{2,1'} & G^{3,1'} \end{bmatrix}$$

$$G_2 = \begin{bmatrix} G^{1,1'} & G^{2,1'} & G^{3,2'} \end{bmatrix}$$

$$G_3 = \begin{bmatrix} G^{1,1'} & G^{2,2'} & G^{3,4'} \end{bmatrix}$$

For comparison of performances, control systems that use G_1–G_8 as the closed-loop gains have the same distance, velocity, and applied force deviations in steady state.

9.6.3 Simulation Results of Trains' Control System Impacted by Packet Drops

The stability of trains' control system in current CBTC systems using the short view strategy is analyzed using Theorem 1 under different packet drop rates. With regard to the short view scheme, Φ_k in Equation 9.28 depends on Θ_k.

$$\Theta_k = \text{Diag}\begin{bmatrix} \gamma_k^0 \theta_k^1 & \gamma_k^1 \theta_k^2 & \cdots & \gamma_k^{n-1} \theta_k^n \end{bmatrix}$$

For the control system of three trains, Θ_k has eight alternatives. Accordingly, there are eight options for Φ_k, $\Phi_k \in \Phi_s, s = 1,\ldots,8$. r_s is the occurrence rate of Φ_s; thus, $\sum_{s=1}^{8} r_s = 1$. For simplicity, we assume that the uplink and downlink packet transmissions are independent and identically Bernoulli distributed random variables.

$$r_s = \left[1 - (1-p)^2 \right]^{3-m} \left[(1-p)^2 \right]^{m}$$

where:
p is the packet drop rate
m is the number of trains that have successfully received LMA

The admissible set of α_s is

$$0.1 \le \alpha_s \le 1, s = 1,2,\dots,7 \text{ and } 1 \le \alpha_8 \le 2.$$

Then we use the LMI toolbox of MATLAB to verify the feasibility of LMIs in Theorem 1. Genetic algorithm (GA) is used to find the maximum decay rate. The fitness function of GA is to find the maximum $\prod_{s=1}^{8} \alpha_s^{r_s}$ throughout the admissible set of α_s that makes all the LMIs feasible.

The exponential stability of system with transmission period $T = 0.3$s under packet drop rate 0.1–0.7 is verified. The maximum decay rate $\alpha^{0.05}$ and the corresponding α_s are given in Table 9.7. It can be seen that the current train's control system in CBTC keeps stable even at a very high packet drop rate, $P(\gamma_k^i = 0) = P(\theta_k^i = 0) = 0.7$. As the drop rate increases, the decay rate of the system decreases.

The performances of T1 and T2 using the Sv scheme under packet drop rate $P(\gamma_k^i = 0) = P(\theta_k^i = 0) = 0.1$ are illustrated in Figure 9.9. The statuses of T1 and T2 are used as the input to T3's controller. The distance, velocity, and applied force deviations of T3 using Sv, Lv_d, or Lv_f scheme under a low packet drop rate $P(\gamma_k^i = 0) = P(\theta_k^i = 0) = 0.01$ are depicted in Figures 9.10 through 9.12, respectively. The deviations of T3 under a much higher packet drop rate $P(\gamma_k^i = 0) = P(\theta_k^i = 0) = 0.1$ are given in Figures 9.13 through 9.15.

It can be found that the Lv_f scheme has the smallest fluctuation of the applied force deviations and the smoothest velocity deviations around the optimal values with a little bit higher distance deviation compared with the Sv and Lv_d schemes. The Lv_d scheme outperforms the Sv scheme in distance, velocity, and applied force deviations.

Table 9.7 LMI Feasibility of the System Using Current Control Scheme ($T = 0.3$ s)

Drop Rate	Decay Rate	α_1	α_2	α_3	α_4	α_5	α_6	α_7	α_8
0.01	1.0331	0.9842	0.9997	0.9927	0.9999	0.9848	0.9993	0.9997	1.9973
0.1	1.0185	0.9938	0.9944	0.9949	0.9999	0.9972	0.9999	0.9972	1.9979
0.2	1.0091	0.9958	0.9999	0.9979	0.9980	0.9968	0.9999	0.9997	1.9984
0.3	1.0040	0.9972	0.9961	0.9998	0.9998	0.9965	0.9998	0.9998	1.9968
0.4	1.0015	0.9997	0.9979	0.9987	0.9998	0.9983	0.9990	0.9999	1.9717
0.5	1.0005	0.9984	0.9997	0.9996	0.9928	0.9995	0.9971	0.9946	1.9927
0.6	1.00008	0.9985	0.9985	0.9999	0.9999	0.9987	0.9955	0.9999	1.9981
0.7	1.00001	0.9999	0.9997	0.9988	0.9933	0.9999	0.9977	0.9986	1.9956

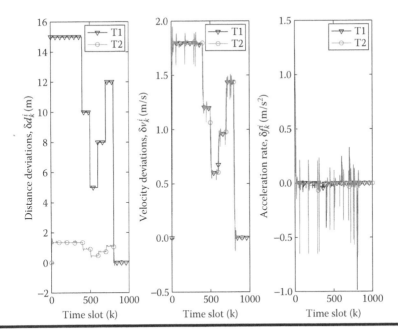

Figure 9.9 Performances of T1 and T2 using the Sv scheme, $T = 0.3$ s, $P(\gamma^i_k = 0) = P(\theta^i_k = 0) = 0.01$.

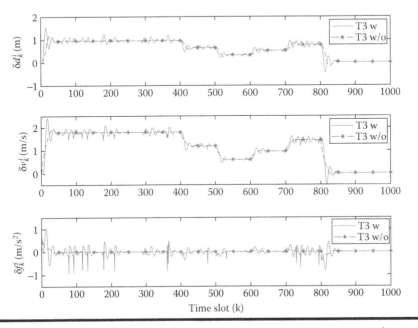

Figure 9.10 Performances of T3 using the Sv scheme, $T = 0.3$ s, $P(\gamma^i_k = 0) = P(\theta^i_k = 0) = 0.01$.

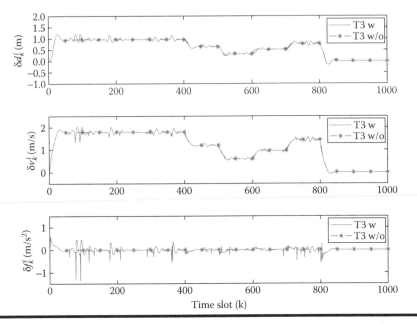

Figure 9.11   Performances of T3 using the Lv_d scheme, $T = 0.3$ s, $P(\gamma_k^i = 0) = P(\theta_k^i = 0) = 0.01$.

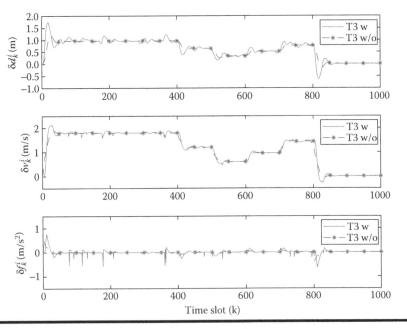

Figure 9.12   Performances of T3 using the Lv_f scheme, $T = 0.3$ s, $P(\gamma_k^i = 0) = P(\theta_k^i = 0) = 0.01$.

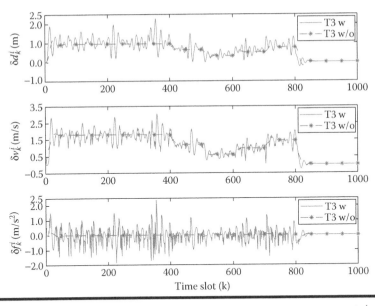

Figure 9.13 **Performances of T3 using the Sv scheme,** $T = 0.3$ s, $P(\gamma_k^i = 0) = P(\theta_k^i = 0) = 0.1$.

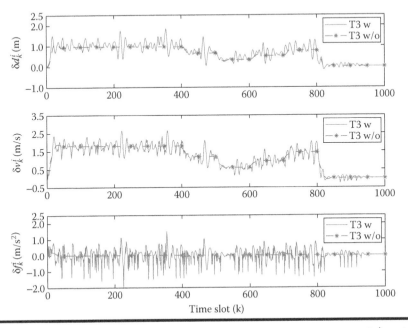

Figure 9.14 **Performances of T3 using the Lv_d scheme,** $T = 0.3$ s, $P(\gamma_k^i = 0) = P(\theta_k^i = 0) = 0.1$.

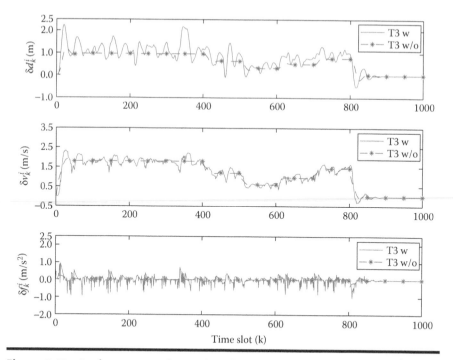

Figure 9.15 Performances of T3 using the Lv_f scheme, $T = 0.3$ s, $P(\gamma_k^i = 0) = P(\theta_k^i = 0) = 0.1$.

The performances of T3 using different control schemes are illustrated in Figure 9.16. It is shown that the Lv_f scheme has a much better applied force and velocity performances at the cost of the distance performance compared with the Sv scheme, which means less energy consumption and better riding comfortability at the expense of line capacity. The Lv_d scheme outperforms the Sv scheme in all kinds of performance measures.

The comparisons of the total cost between the proposed Lv_f/Lv_d scheme and the Sv scheme with different coefficients are given in Figure 9.17. It can be found that both the Lv_f and Lv_d schemes outperform the Sv scheme.

9.7 Conclusion

CBTC systems use wireless networks to transmit trains' status and control data. In this chapter, the issues related to trains' control in CBTC systems with lossy wireless networks were studied. The system was modeled as an NCS with packet drops in both uplink and downlink transmissions. We studied packet drops introduced by random transmission errors and handovers. The packet drop rate was formulated and related to handovers in CBTC. The analytical results on the rate of packet

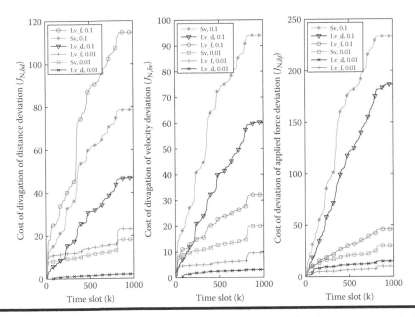

Figure 9.16 Cost of T3 using the Sv, Lv_f, and Lv_d schemes, respectively, $T = 0.3$ s, $P(\gamma_k^i = 0) = P(\theta_k^i = 0) = 0.1, 0.01$.

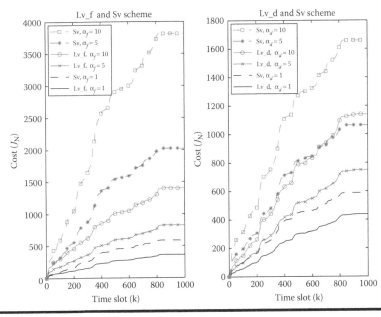

Figure 9.17 Total cost of T3 using the Sv, Lv_f, and Lv_d schemes, respectively, $T = 0.3$ s, $P(\gamma_k^i = 0) = P(\theta_k^i = 0) = 0.1$.

drops introduced by handover were proved to be in line with the field test results obtained from business operating CBTC lines. The effects of packet drops on the stability and performances of trains' control systems were analyzed.

We proposed two novel schemes to improve performances of trains' control system under packet drops. Each train uses different combinations of the status or estimated status of foregoing trains to generate control commands through selecting different closed-loop gains to minimize divagation of the applied force or distance deviations from the optimal values. Extensive simulation results of systems with currently adopted and proposed schemes under different packet drop rates were presented. The results showed that the proposed schemes outperform the currently used scheme in CBTC systems. They can provide less energy consumption, better riding comfortability, and higher line capacity.

References

1. R.D. Pascoe and T.N. Eichorn. What is communication-based train control? *IEEE Veh. Technol. Mag.*, 4(4):16–21, 2009.
2. IEEE. *IEEE Std 1474.1-2004: IEEE Standard for Communications-Based Train Control (CBTC) Performance and Functional Requirements.*
3. M. Aguado, E. Jacob, P. Saiz, J.J. Unzilla, M.V. Higuero, and J. Matias. Railway signaling systems and new trends in wireless data communication. In *Proc. IEEE VTC2005-FALL*, Dallas, TX, pages 1333–1336, September 2005.
4. R. Lardennois. Wireless communication for signaling in mass transit, Siemens Transportation Systems 2003.
5. E. Kuun and W. Richard. Open standards for CBTC and CCTV radio-based communication. *Alcatel Telecommun. Rev.*, (2):243–252, 2004.
6. P. Howlett. An optimal strategy for the control of a train. *J. Aust. Math. Soc. Ser. B-Appl. Math.*, 31(Part 4):454–471, 1990.
7. P. Howlett. Optimal strategies for the control of a train. *Automatica*, 32(4):519–532, 1996.
8. P. Howlett. The optimal control of a train. *Ann. Oper. Res.*, 98:65–87, 2000.
9. P.G. Howlett and J. Cheng. Optimal driving strategies for a train on a track with continuously varying gradient. *J. Aust. Math. Soc. Ser. B-Appl. Math.*, 38(Part 3): 388–410, 1997.
10. P.G. Howlett, P.J. Pudney, and X. Vu. Local energy minimization in optimal train control. *Automatica*, 45(11):2692–2698, 2009.
11. J. Cheng and P. Howlett. A note on the calculation of optimal strategies for the minimization of fuel consumption in the control of trains. *IEEE Trans. Auto. Contr.*, 38(11):1730–1734, 1993.
12. L. Zhu, F. R. Yu, B. Ning, and T. Tang. Cross-layer design for video transmissions in metro passenger information systems. *IEEE Trans. Veh. Technol.*, 60(3):1171–1181, 2011.
13. L. Zhu, F.R. Yu, B. Ning, and T. Tang. Handoff management in communication-based train control networks using stream control transmission protocol and IEEE 802.11p WLANs. *Eurasip. J. Wirel. Comm.*, 2012(1):1–16, 2012.

14. L. Zhu, F.R. Yu, B. Ning, and T. Tang. Handoff performance improvements in MIMO-enabled communication-based train control systems. *IEEE Trans. Intell. Transp.*, 13(2):582–593, 2012.

15. Z. Li, F. Richard Yu, and B. Ning. Service availability analysis in communication-based train control systems using wireless local area networks. *Wirel. Commun. Mob. Comp.*, 15:16–19, 2012.

16. J. Tang and X. Zhang. Cross-layer modeling for quality of service guarantees over wireless links. *IEEE Trans. Wirel. Commun.*, 6(12):4504–4512, 2007.

17. J. Tang and X. Zhang. Cross-layer resource allocation over wireless relay networks for quality of service provisioning. *IEEE J. Sel. Area. Comm.*, 25(4):645–656, 2007.

18. A.L. Toledo and X.D. Wang. TCP performance over wireless MIMO channels with ARQ and packet combining. *IEEE Trans. Mobile. Comput.*, 5(3):208–223, 2006.

19. F. Richard Yu, B. Sun, V. Krishnamurthy, and S. Ali. Application layer QoS optimization for multimedia transmission over cognitive radio networks. *ACM/Springer Wireless Networks*, 17(2):371–383, 2011.

20. F. Yu and V. Krishnamurthy. Effective bandwidth of multimedia traffic in packet wireless CDMA networks with lmmse receivers: A cross-layer perspective. *IEEE Trans. Wirel. Commun.*, 5(3):525–530, 2006.

21. L. Zhu, F. R. Yu, B. Ning, and T. Tang. Cross-layer handoff design in MIMO-enabled WLANs for communication-based train control (CBTC) systems. *IEEE J. Sel. Area. Comm.*, 30(4):719–728, 2012.

22. A. Rabello and A. Bhaya. Stability of asynchronous dynamical systems with rates constraints and applications. *IEE Proc. Control Theory Appl.*, 150(5):546–550, 2003.

23. J.P. Hespanha, P. Naghshtabrizi, and Y. Xu. A survey of recent results in networked control systems. *Proc. IEEE*, 95(1):138–162, 2007.

24. S. Hu and W.-Y. Yan. Stability robustness of networked control systems with respect to packet loss. *Automatica*, 43(7):1243–1248, 2007.

25. O.C. Imer, S. Yueksel, and T. Basar. Optimal control of LTI systems over unreliable communication links. *Automatica*, 42(9):1429–1439, 2006.

26. P. Seiler and R. Sengupta. Analysis of communication losses in vehicle control problems. In *Proc. Amer. Contr. Conf.*, volume 2, Arlington, VA, pages 1491–1496, 2001.

27. F. Wang and D. Liu. *Networked control systems*. Springer, New York, 1st edition, 2008.

28. J. Wu and T. Chen. Design of networked control systems with packet dropouts. *IEEE Trans. Automat. Contr.*, 52(7):1314–1319, 2007.

29. J.K. Hedrick, M. Tomizuka, and P. Varaiya. Control issues in automated highway systems. *IEEE Control Syst. Mag.*, 14(6):21–32, 1994.

30. R.H. Middleton and J.H. Braslavsky. String instability in classes of linear time invariant formation control with limited communication range. *IEEE Trans. Automat. Contr.*, 55(7):1519–1530, 2010.

31. W. Levine and M. Athans. On the optimal error regulation of a string of moving vehicles. *IEEE Trans. Automat. Contr.*, 11(3):355–361, 1966.

32. IEEE. *IEEE Std 802.11-2007: IEEE Standard for Information Technology-Telecommunications and Information Exchange between Systems-Local and Metropolitan Area Networks-Specific Requirements*. Part 11 Wireless LAN medium access control (MAC) and physical layer (PHY) specifications.

33. M.S. Iacobucci, G. Paris, D. Simboli, and G. Ziti. Analysis and performance evaluation of wireless LAN handover. In *Proc. 2nd ISWCS*, Siena, Italy, pages 337–341, September 2005.

34. A. Mishra, M. Shin, and W. Arbaugh. An empirical analysis of the IEEE 802.11 MAC layer handoff process. *Comput. Commun. Rev.*, 33(2):93–102, 2003.

35. J. Montavont, N. Montavont, and T. Noel. Enhanced schemes for L2 handover in IEEE 802.11 networks and their evaluations. In *Proc. IEEE PIMRC*, volume 3, Berlin, Germany, pages 1429–1434, September 2005.

36. L. Zhu, Y. Zhang, B. Ning, and H. Jiang. Train-ground communication in CBTC based on 802.11b: Design and performance research. In *Proc. CMC*, volume 2, Yunnan, China, pages 368–372, January 2009.

37. D.B. Green and A.S. Obaidat. An accurate line of sight propagation performance model for ad-hoc 802.11 wireless LAN (WLAN) devices. In *Proc. IEEE ICC*, volume 5, New York, pages 3424–3428, May 2002.

38. Gordon L. Stuber. *Principles of Mobile Communication*. Springer, 3rd edition, 2011.

39. A.F. Molisch. *Wireless communications*. John Wiley & Sons, Hoboken, NJ, 2nd edition, 2011.

40. R.V. Dukkipati. *Control Systems*. Alpha Science International Ltd. Oxford, UK.

41. C.L. Phillips and R.D. Harbor. *Feedback Control Systems*. Tom Robbins, Upper Saddle River, NJ, 4th edition, 2000.

Chapter 10

Cognitive Control for Communications-Based Train Control Systems

Hongwei Wang and F. Richard Yu

Contents

10.1 Introduction 214
10.2 Overview of Cognitive Control 216
    10.2.1 Cognitive Control Approach to CBTC Systems 217
10.3 Cognitive Control 218
10.4 Formulation of Cognitive Control Approach to CBTC Systems............221
    10.4.1 Train Control Model..221
    10.4.2 Channel Model in MIMO-Enabled WLANs..........................223
    10.4.3 *Q*-Learning in the Cognitive Control Approach......................225
        10.4.3.1 System States and Actions 225
        10.4.3.2 Reward Function..226
10.5 Simulation Results and Discussions..231
    10.5.1 Parameters of Train Dynamics 231
    10.5.2 Parameters of the ATO...232
    10.5.3 Parameters of the Wireless Channel 233
    10.5.4 Simulation Results and Discussions..233
10.6 Conclusion 242
References 243

10.1 Introduction

Urban rail transit systems have developed rapidly around the world in the recent past. Due to the huge urban traffic pressure, improving the efficiency of urban rail transit systems is in high demand. As a key subsystem of urban rail transit systems, communications-based train control (CBTC) is an automated train control system using train–ground communications to ensure the safe and efficient operation of rail vehicles [1]. CBTC can improve the utilization of railway network infrastructure and enhance the level of service offered to customers [2].

As urban rail transit systems are built in a variety of environments (e.g., underground tunnels, viaducts, etc.), there are different wireless network configurations and propagation schemes. For tunnels, the free space is generally adopted as the propagation medium. However, the leaky coaxial cable is also an option, such as Tianjin Subway lines 1 and 2 built by Bombardier. For the viaduct scenarios, leaky rectangular waveguide is a popular approach, as it can provide higher performance and stronger anti-interference ability than the free space [3]. In addition, due to the available commercial off-the-shelf (COTS) equipment, wireless local area networks (WLANs) are often adopted as the main method of train–ground communications for CBTC systems [4].

Building a train control system over wireless networks is a challenging task. Due to unreliable wireless communications and train mobility, the train control performance can be significantly affected by wireless networks [5]. Because CBTC systems are safety critical, trains usually run according to the front train's state, including velocity and position. When a wireless network brings large communication latency caused by unreliable wireless communications or handoffs, the current train may not be able to obtain the accurate state information of the front train, which would severely affect train operation efficiency, or even cause train emergency brake.

The performance issues in railway environments have attracted a lot of interest recently. A fast handoff algorithm suitable for passenger lines is proposed in [6] by setting a neighboring cell list to facilitate handoff operations. In [7], a novel handoff scheme based on on-vehicle antennas is introduced. The authors of [8] propose a cross-layer handoff design in for multiple-input and multiple-output (MIMO)-enabled WLANs in CBTC systems. In [9], energy-efficient train control schemes are studied in CBTC systems.

Although these above excellent works have been done to study CBTC systems from both *train–ground communication* and *train control* perspectives, these two important areas have traditionally been addressed separately in the CBTC literature. However, as shown in the following, it is necessary to jointly consider these two advanced technologies together to enhance the level of safety and services in CBTC systems. The motivation behind our work is based on the following observations.

▪ Most existing CBTC systems are based on the COTS equipment, in which traditional design criteria (e.g., network capacity) are used in the design and configuration of these systems. Although traditional design criteria are suitable for commercial networks (e.g., hot spots for Internet access), they may not be suitable for CBTC systems due to their specific characteristics, such as stringent requirements for communication availability and latency.

■ Recent studies in cross-layer design show that maximizing network capacity does not necessarily benefit the application layer [10,11], which is train control in CBTC systems. From a CBTC perspective, the performance of train control is more important than that of other layers.

■ Most train control schemes in existing CBTC systems assume perfect train–ground communication. However, random transmission delays and packet drops are inevitable in train–ground communications, which will significantly affect train control performance in CBTC systems [12].

In this chapter, we propose to jointly study train–ground communication and train control so as to improve CBTC performance. The distinct features of this work are as follows:

■ With recent advances in cognitive dynamic systems [13,14], we take a *cognitive control* approach to CBTC systems considering both train–ground communication and train control. Recently, cognitive dynamic systems have emerged as a new engineering discipline, which builds on ideas in statistical signal processing, stochastic control, and information theory. This discipline has been successfully used in the design of dynamic systems (e.g., cognitive radio and cognitive radar) with efficiency, effectiveness, and robustness being the hallmarks of performance [13].

■ In our cognitive control approach, the notion of *information gap* [14] is adopted to quantitatively describe the effects of train–ground communication on train control performance. Specifically, as train–ground communication is used to exchange information between the train and the control center, packet delay and drop lead to information gap, which is the difference between the actual state and the observed state of the train.

■ Unlike the existing works that use network capacity as the design measure, in this chapter, linear quadratic cost for the train control performance in CBTC systems is considered in the performance measure. Reinforcement learning (RL) is applied to obtain the optimal policy based on the performance measure, which includes linear quadratic cost and information gap.

■ The wireless channel is modeled as finite-state Markov chains with multiple state transition probability matrices, which can demonstrate the characteristics of both large-scale fading and small-scale fading. The channel state transition probability matrices are derived from real field measurement results.

■ Simulation results show that the proposed cognitive control approach can significantly improve the train control performance in CBTC systems.

The rest of this chapter is organized as follows: Section 10.2 gives the introduction of cognitive control. Section 10.3 describes the cognitive dynamic systems and the RL. Section 10.4 discusses the models. Section 10.5 presents the optimal solutions, simulation results, and some discussions. Finally, Section 10.6 concludes the chapter.

10.2 Overview of Cognitive Control

Cognitive control was originally developed in neuroscience and psychology (e.g., [15]). Recently, it has emerged as a new engineering discipline [14]. The basic diagram of a cognitive control system is shown in Figure 10.1.

Compared with other control methods, such as adaptive control [16] and neuro-control [17], cognitive control has the following advantages: There is no memory block in the adaptive controller, which reduces the ability of learning. The neuro-controller lacks intelligence, which is distributed throughout the cognitive dynamic system and can make the system in an orderly fashion. The concept of cognitive control has been successfully applied in cognitive radar systems and cognitive radio systems [18]. According to Figure 10.1, the feedback information

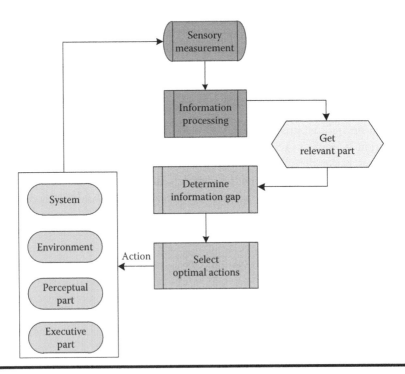

Figure 10.1 Basic schematic structure of a cognitive control system.

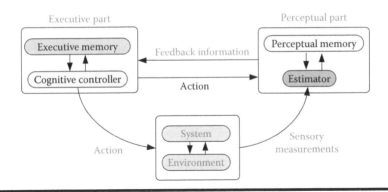

Figure 10.2 Basic procedure of a cognitive control system.

plays the key role in a cognitive dynamic system, and cognitive control describes a control system from the information flow perspective. The feedback information obtained by the perceptual part is partitioned into *relevant information* and *redundant information*. However, the required information is used to make correct decisions, which is called *sufficient information* in cognitive control. The information gap is defined as the difference between sufficient information and relevant information obtained from the measurements, and the basic procedure of cognitive control is shown in Figure 10.2. The goal of cognitive control is to decrease the information gap.

10.2.1 Cognitive Control Approach to CBTC Systems

Cognitive control can be applied in CBTC systems to jointly consider both train–ground communication and train control. The information gap of CBTC systems can be defined according to the train control procedure, in which the current train is controlled according to the information of the front train. In CBTC systems, the train is controlled by the command generated from automatic train operator (ATO). At each control cycle, ATO determines the decision of the train operation according to the automatic train protection (ATP) emergency braking profile and the location of movement authority (MA), which means that the ATO model is also necessary. As cognitive control improves the train efficiency through selecting the optimized communication policy, the channel model of CBTC operation environment is important, where the accuracy of channel model directly affects the accuracy of control. Therefore, adopting cognitive control in CBTC systems needs the dynamic model, the ATO model, and the channel model.

Based on the principles of CBTC train control, whether the MA is transmitted timely decides the performance of whole CBTC system. MA is the basis for ATO and ATP decisions, which is generated from zone controller (ZC) according to the state information of the front train. An MA is generally defined as a physical point on the track. It is the nearest potential obstacle in front of the train, such as the tail of the front train. In some scenarios, when ATP calculates an emergency braking profile and ATO calculates an operating speed/distance profile based on MA, MA is often taken as the distance from the front end of the current train to the tail of the front train.

In CBTC systems, the current train needs the information of the front train to control acceleration/deceleration at every communication cycle. If ZC can send the accurate information to the current train, which means that the current train can get sufficient information, the current train can make correct decisions. In CBTC systems, ZC transmits an MA to the current train according to the information sent from the front train. An MA is generally defined as a physical point on the track. It is the nearest potential obstacle in front of the train, such as the tail of the front train. In some scenarios, when ATP calculates an emergency braking profile and ATO calculates an operating speed/distance profile based on MA, MA is often taken as the distance from the front end of the current train to the tail of the front train.

As mentioned earlier, due to unreliable wireless communications and handoffs, the information included in the received MA by the current train may not exactly describe the state of the front train. As a result, we can see that the information gap in CBTC systems is the difference between the derived state of the front train from the received MA sent by ZC and the actual state of the front train.

In this chapter, we take a cognitive control approach to CBTC systems considering both train–ground communication and train control, and information gap is used to quantitatively describe the effects of train–ground communication on train control performance.

10.3 Cognitive Control

In this section, we describe cognitive control in detail. The cost function is defined. Then, we present RL to derive the optimal policy in cognitive control.

For a cognitive control system shown in Figure 10.1, the perceptual part contains the estimator and the perceptual memory, where the estimator is to obtain the available information from the sensory measurements results and the perceptual memory can process the information to get the relevant part. The cognitive controller of the executive part makes corresponding decisions based on the knowledge in the executive memory according to the feedback information from the perceptual part. Based on the knowledge in the executive memory, the cognitive controller selects the optimal action, which has influence on the system itself or the environment. When it acts on the system, the sensors or the actuators may be reconfigured. When it acts on the environment, the perception process could be indirectly affected. In fact, the key of cognitive control is that the cognitive actions might be a part of physical actions (state-control actions). In other words, a physical action is applied and the goal is to decrease the information gap. For example, when there is a quadratic optimal controller, the cost function is

$$J = (x - \tilde{x})^T Q(x - \tilde{x}) + u^T Ru \qquad (10.1)$$

where:
$\tilde{x}$ is the desired state of the system
x is the actual state of the system
The matrices Q and R are applied as the desired weights for systems's state and control

The objective is to minimize the cost function.

Moreover, cognitive control adds another term about information gap to Equation 10.1. The resulting cost function can be formulated as [14]

$$J = (x - \tilde{x})^T Q(x - \tilde{x}) + u^T Ru + \beta G \qquad (10.2)$$

where:
G is the information gap
β is a scalar

In cognitive control, the cognitive actions are of greatest concern. The actions are determined through the implementation of RL. RL is the process by which the agent learns an approximately optimal policy through trial-and-error interactions with the environment. At each communication cycle, RL can determine the cognitive action to decrease the information gap according to the reward. As a result, the objective of RL is to find a policy that is updated by rewards provided by the environment, which means minimizing the cumulative amount of cost over a long run [19].

In the RL model depicted in Figure 10.3, a learning agent selects an action acting on the system or environment according to the current system state and the current environment state. When the new state comes, the agent gathers information about the new state and calculates the immediate reward and the time that the state transition costs. Then, based on the information and an algorithm, the agent can update the knowledge base and select the next action. The agent continues to improve its performance with the process repeated.

Let $S = \{s_1, s_2, ..., s_n\}$ be the set of system states and $A = \{a_1, a_2, ..., a_m\}$ be the set of actions. According to the current state $s_k \in S$, the RL agent interacts with the environment and chooses the action $a_k \in A$. Then there is a state transition, and the

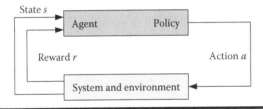

Figure 10.3   RL model.

new state is $s_{k+1} \in S$ based on the state transition probability. The immediate reward is given and the process is repeated.

As Q-learning is one of the most popular RL algorithms, it is adopted in this chapter to find an optimal policy for the learning agent. The Q-learning algorithm has a Q-function that calculates the *quality* of a state–action combination, $Q : S \times A \rightarrow \mathbf{R}$. In fact, Q-function is an evaluation function $Q(s, a)$ so that its value is the maximum discounted cumulative reward $r(s, a)$ that can be achieved starting from state s and applying action a as the first action. In other words, the value of Q is the reward received immediately upon executing action a from state s, plus the value of following the optimal policy (discounted by η) thereafter [20].

$$Q(s,a) \equiv r(s,a) + \eta\{V^*[\delta(s,a)]\} \tag{10.3}$$

where:

$r(s, a)$ is the immediate reward

$\delta(s, a)$ denotes the state resulting from applying action a to state s

$V^*(s)$ gives the maximum discounted cumulative reward that the agent can obtain starting from state s, and it is defined as $V(s_k) = \sum_{i=0}^{\infty} \eta^i r_{k+i}$, where $0 \leq \eta < 1$

As there are state transition probabilities, Equation 10.3 can be rewritten as follows:

$$Q^\pi(s,a) = r(s,a) + \eta \sum_{s' \in S} P_{ss'}(a)V^\pi(s') \tag{10.4}$$

where:

$s' = \delta(s, a)$

$p_{ss'}$ is the transition probability from state s to state s' when applying action a

Now, we define a policy $\pi(s) \in A$, and the optimal policy is denoted as $\pi^*(s)$. Then, we get

$$Q^*(s,a) = Q^{\pi(s,a)^*} = r(s,a) + \eta \sum_{s' \in S} P_{ss'}(a)V^{\pi^*}(s') \tag{10.5}$$

where:

$V^{\pi^*}(s) = V^*(s)$

Notice the close relationship between Q and V^*: $V^*(s) = \max_{a \in A} Q^*(s, a)$. Then, Equation 10.5 can be written as

$$Q^*(s,a) = r(s,a) + \eta \sum_{s' \in S} P_{ss'}(a)\max_{a' \in A} Q^*(s',a') \tag{10.6}$$

As a result, the optimal policy can be defined as $\pi^*(s) = \operatorname{argmax}_{a \in A} Q^*(s, a)$, which means the Q-learning rules can be determined.

10.4 Formulation of Cognitive Control Approach to CBTC Systems

In the section, we illustrate each part of the cognitive control model of CBTC systems, including the control model, wireless channel model, and Q-learning parameters.

ATP is the safety guard of CBTC systems. For each communication cycle, ATP calculates the permitted maximum speed for the train base on MA. According to the permitted speed, the limited speed of the line, and the state of the train, ATO determines to accelerate or decelerate. When MA is delayed due to the communication latency, ATO system finds that the speed might be beyond the permitted speed calculated by ATP, and ATO will bring the deceleration command to force the train to brake or stop in order to keep safe. Therefore, ATP has the higher safety level, which generates the basis for the ATO decision. When cognitive control is applied in CBTC systems, ATP should be considered as the constraint in the control procedure.

The general structure of cognitive control consists of three parts: the executive part including a cognitive controller, the perceptual part, and the practical environment including the system, as shown in Figure 10.1. We adopt cognitive control to model a CBTC system, where ZC can be taken as the perceptual part and send MA to the current train. According to the state of the system, the environmental conditions, and the output of ATO, the cognitive controller can control the mobile station (MS) in the current train to perform cognitive actions to decrease the information gap, such as the trade-off among MIMO diversity gain, multiplexing gain, and the handoff decisions. These actions have impacts on performance of wireless communications, such as signal-to-noise (SNR) and data rate. Similarly, the wireless channel model plays an important role in the perceptual memory, while the Q-learning can get the knowledge stored in the executive memory. Hence, the structure of the cognitive control approach is illustrated in Figure 10.4. Next, we need to determine the train control model, which generates the inputs of the cognitive controller.

10.4.1 Train Control Model

Generally, ATO receives the MA from ZC, calculates the operation speed curve, and determines the acceleration (deceleration) of the next communication cycle for the train. According to Newton's Second Law, the state of the train can be updated. In this chapter, we assume that the train controller is linear time invariant in discrete time. Then, the dynamic model of the train can be shown as follows:

$$x_{k+1} = Ax_k + Bu_k + Cw_k$$

$$z_k = C_1 x_k + D_1 u_k \tag{10.7}$$

$$y_k = C_2 x_k$$

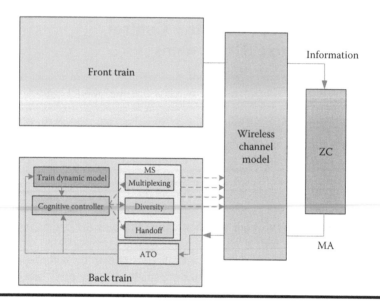

Figure 10.4 Schematic structure of the cognitive control approach to CBTC systems.

where:

x_k is the state of the train in the kth communication cycle including the position and the velocity

u_k is the input of the train in the kth communication cycle that is determined by the ATO model

w_k is the resistance related to the train in the kth communication cycle

z_k is the regulated output in the kth communication cycle

y_k is the measured output in the kth communication cycle

Hence, the ATO model has the following state space model:

$$x^c_{k+1} = A^c x^c_k + B^c y_k$$
$$u_k = C^c f(x^c_k)$$

(10.8)

where:

y_k is the input of ATO, which includes the states of the two trains

Function $f(\cdot)$ converts the information of two trains into the parameters that can be used to directly calculate the acceleration (deceleration)

The linear quadratic cost is taken as our performance measure of the train control in this chapter. The general expression is

$$J = \sum_{k \to \infty} [(x_k - \tilde{x}_k)^T Q(x_k - \tilde{x}_k) + u_k^T R u_k] \tag{10.9}$$

where:

$\tilde{x}_k$ is the desired state of the train

x_k is the actual state of the train

Q is semipositive definite

R is positive definite

In fact, the first term is the state tracking error and the second term is the control magnitude, which is related to energy consumption.

When the cognitive control is applied, the resulting cost function can be obtained according to Equation 10.2

$$J = \sum_{k \to \infty} [(x_k - \tilde{x}_k)^T Q(x_k - \tilde{x}_k) + u_k^T R u_k + \beta G_k] \tag{10.10}$$

where:

$\tilde{x}_k$ is the information gap in the kth communication cycle

β is a scalar

The objective of the optimal control is to realize efficient train operations and the minimum energy consumption through decreasing the information gap.

10.4.2 Channel Model in MIMO-Enabled WLANs

In order to optimize the performance of CBTC systems, we build a finite-state Markov channel (FSMC) model based on the real field measurements. FSMC models have been widely accepted in the literature as an effective approach to characterize wireless channels, including high-speed railway channels [21], satellite channels [22], and Rayleigh fading channels [23]. In FSMC models, the SNR range of the received signal can be partitioned into nonoverlapping levels. Then the received SNR can be modeled as a random variable evolving according to an FSMC with state transition probabilities, which can be obtained from real field channel measurements.

Due to the effect of large-scale fading, the amplitude of SNR depends on the distance between the transmitter and the receiver. It is obvious that the SNR is usually high when the receiver is close to the transmitter; whereas it is low when the receiver is far away from the transmitter. As a result, the transition probability from the high channel state to the low channel state is different when the receiver is near or far away from the transmitter, which means that the Markov state transition probability is related to the location of the receiver. Therefore, only one state transition probability matrix, which is independent of the location of the receiver, may not accurately model the channels.

As a result, we divide the communication coverage of one access point (AP) into L intervals. For each interval, we use Lloyd–Max method to partition the SNR amplitude into several levels, which are nonuniformly distributed. The non-uniformed partitioning can be useful to obtain more accurate estimates of system performance measures [24]. Specifically, $\mathbf{P}^l$, $l \in \{1,2,...,L\}$ is the state transition probability matrix corresponding to the lth interval, and the relationship between the transition probability and the location of the receiver can be built. Then, $p_{n,j}^l$ is the state transition probability from state s_n^l to state s_j^l in the lth interval. As a result, the state probabilities and the state transition probabilities can be defined as follows:

$$p_n^l = P_r^l \{\gamma_k^l = s_n^l\}$$
$$p_{n,j}^l = P_r^l \{\gamma_{k+1}^l = s_n^l \mid \gamma_k^l = s_j^l\}$$
$$p_{n,j}^l = 0, \text{if} \mid n - j \mid > 1 \tag{10.11}$$
$$\sum_{j=1}^{N} p_{n,j}^l = 1, \forall n \in \{1,2,3,...,N\}$$

where:

p_n^l is the probability of being in state n in the lth interval

γ_k^l is the channel state in time slot k in the lth interval

In this chapter, a MIMO-enabled WLAN is used in CBTC systems. A MIMO system can provide two types of gains: diversity gain and spatial multiplexing gain, which may not be obtained simultaneously. There is a fundamental trade-off between diversity gain and spatial multiplexing gain [25]: higher spatial multiplexing gain comes at the price of sacrificing diversity. We can achieve the optimal diversity gain $d^*(r)$ with long enough block length as follows:

$$d^*(r) = (m_t - r)(n_r - r) \tag{10.12}$$

where:

r is the spatial multiplexing gain, which is an integer

m_t and n_r are the numbers of transmitting and receiving antennas, respectively, which correspond to the numbers of the AP antennas and the MS antennas

In particular, $d_{max}^* = m_t n_r$ and $r_{max}^* = \min\{m_t, n_r\}$.

With a multiplexing gain r, the data rate $C(r)$ and the bit error rate (BER) probability $b(r)$ can be approximated as [26]

$$C(r) = k_c r \log_2(\gamma)$$
$$b(r) = k_p \gamma^{-d(r)} \tag{10.13}$$

where:

k_c and k_p are positive constants for different coding schemes

As a result, the data rate and the BER can be obtained when the multiplexing gain and the diversity gain are obtained through Equation 10.12, which can be used to derive the performance of the higher layers.

10.4.3 Q-Learning in the Cognitive Control Approach

The overall systems can be considered as a discrete-time event system. As the wireless channel is modeled as an FSMC channel, each channel state can only transit to the adjacent channel states. In order to utilize Q-learning algorithm, the system states, actions, and rewards should be identified. The RL model in the cognitive control approach is shown in Figure 10.5.

10.4.3.1 System States and Actions

In the Q-learning model, the cognitive controller on the train should decide if a handoff procedure happens from the connected AP to the other available AP at each communication cycle, where we assume that the place occupied by a train can only be covered by two successive APs. As mentioned earlier, the multiplexing gain

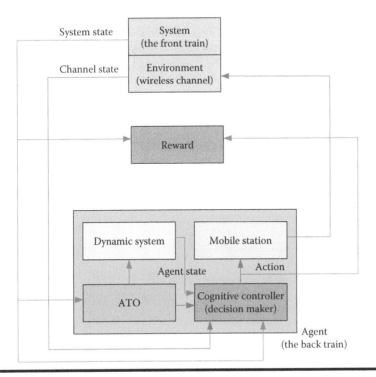

Figure 10.5   RL model in the cognitive control approach.

needs to be determined to improve the channel conditions. The handoff action is denoted as a_k^h at the kth communication cycle and the multiplexing gain action is a_k^m. Then, the current action $a_k \in A$ is $a_k = \{a_k^h, a_k^m\}$.

Corresponding to the actions, the current states should indicate the physical layer of wireless communications and the handoff procedure, such as the SNR levels of two adjacent APs. Because our FSMC model is related to the distance between the transmitter and the receiver, the channel state is given as γ_{1k}^l and γ_{2k}^l, which are the SNR levels of two successive APs in the lth interval. As a result, the current state $s_k \in S$ is denoted as $s_k = \{\gamma_{1k}^l, \gamma_{2k}^l, \text{ID}\}$, where ID is the identification number of the current associated AP. When ID changes, the handoff procedure happens.

10.4.3.2 Reward Function

When each action is taken, the system earns deterministic reward, which is used to demonstrate the effects of the action on the system. In our Q-learning model, the linear quadratic cost should be minimized according to the optimization objective shown in Equation 10.10, which includes the guidance trajectory tracking error, the control magnitude, and the information gap. As a result, we take the reciprocal of the linear quadratic cost as the reward function. Hence, the communication delay may affect the performance of the tracking, which can cause frequent acceleration and deceleration and increase the energy assumption. There are two kinds of communication delay: delay with handoffs (handoff latency) and delay without handoffs. Therefore, we should first study the communication delay without handoffs.

IEEE 802.11g WLANs are applied as the main method of the train–ground communication in CBTC systems, where carrier sense multiple access/collision avoidance (CSMA/CA) is used in the media access control (MAC) layer. When the train is running with high speed, the wireless channel can be affected due to the Doppler frequency shift, reflections, and other factors. The performance of the physical layer will be decreased, such as packet loss rate (PLR). The packet loss will bring retransmission of data packets according to the automatic repeat request scheme with CSMA/CA, which can cause time delay. According to Equation 10.13, the BER can be obtained. Then, the corresponding frame error rate (FER) is derived.

$$f(r) = 1 - [1 - b(r)]^{L_f} \tag{10.14}$$

where:

L_f is the length of one frame whose unit is bit

For a given FER, the PLR can be obtained.

$$p(r) = f(r)^{\alpha-1}[1 - f(r)] \tag{10.15}$$

where:

α is the times of packet retransmissions

Generally, there are two factors that can cause the packet loss: (1) the packet collision and (2) the channel error. In the CBTC scenarios, there may not be many trains occupying one AP's coverage. As a result, there may not be packet collisions. Then, the only factor we are concerned about in this chapter is the channel error that can lead to packet retransmissions. First of all, we can derive that the time delay without retransmission ($\alpha = 0$) is caused by the random access scheme according to the CSMA/CA method in 802.11g [27].

$$T_r^0 = aDifsTime + aDataTime + aSifsTime$$
$$+anACKTime + aPropTime \tag{10.16}$$

where:

$aDifsTime$ is the period of distributed interframe space (DIFS)
$aDataTime$ is the time for the transmitter to send a data frame and determined by the ratio of the length of the frame L_f and the current data rate $C(r)$
$aSifsTime$ is the period of short interframe space (SIFS)
$anACKTime$ is the time for the transmitter to send an acknowledge frame
$aPropTime$ is the propagation time

When the packet loss happens, the retransmission scheme is triggered with a backoff time, which is uniformly distributed in the range $[0, CW - 1]$, where CW is the contention window. For the first backoff, CW is initialized as CW_{min}. Then, each transmission attempt will double CW until it reaches CW_{max}. In the 802.11 standard [27], $CW = \{16, 32, 64, 128, 256, 512, 1024\}$ and $CW_{min} = 16$, $CW_{max} = 1024$. Then the general expression of CW can be denoted as

$$CW_\alpha = \begin{cases} 2^{4+\alpha}, 0 \le \alpha \le 6 \\ 2^{10}, \alpha > 6 \end{cases} \tag{10.17}$$

where α is the times of retransmissions.

As a result, the backoff time can be defined as

$$BackoffTime_\alpha = Random([0, CW_\alpha - 1]) \times aSlotTime \tag{10.18}$$

where:

$aSlotTime$ is a constant time corresponding to IEEE standard of 802.11g [27]

Hence, we can derive the time delay caused by retransmissions.

$$T_r^\alpha = \sum_{i=0}^{\alpha} (BackoffTime_i) + aPropTimes$$
$$+\alpha(aDifsTime + aDataTime + aSifsTime) \tag{10.19}$$

Combining Equations 10.14 and 10.19, the expectation of time delay without handoff can be derived due to the quality of the wireless channel.

$$T_d^e = \sum_{i=0}^{\alpha} f(r)^i [1 - f(r)] T_r^i \tag{10.20}$$

Now we should consider the handoff latency, which generally brings more effects on the performance of train control. The handoff procedure commonly consists of three steps: scanning, authentication, and association, ended with one deauthentication frame sent from the MS, as shown in Figure 10.6. There are seven frames transmitted before a complete handoff procedure is finished: probe request frame, probe response frame, authentication request frame, authentication response frame, association request frame, association response frame, and deauthentication

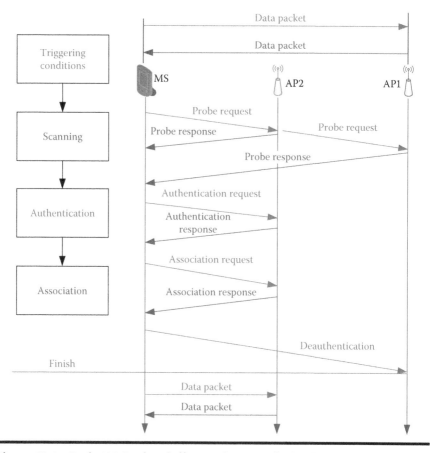

Figure 10.6 Basic WLAN handoff procedure. At the beginning, MS communicates with AP1. After the handoff procedure finished with one deauthentication frame sent from the MS, MS communicates with AP2.

frame. As a result, the expectation of the time that a complete handoff procedure costs is about $7T_d^e$. At the scanning step, if it happens that the broadcasting probe request frame is successfully sent or there are some probe response frames received by the MS, the MS will take the probe period time denoted as T_{period} to reach the authentication step. However, each of the other five frames cannot be transmitted and a new handoff procedure will be restarted, which will cost long time denoted by T_{new} for the MS to start a new handoff procedure. As a result, the expectation of handoff latency with handoff can be derived as

$$
\begin{aligned}
T_h^e = (1 - p(r))^2 &\{(1 - p(r))^4 (7T_d^e + T_{\text{period}}) \\
&+[1 - (1 - p(r))^4](7T_d^e + T_{\text{period}} + T_{\text{new}})\} \\
&+[1 - (1 - p(r))^2]\{(1 - p(r))^4 (7T_d^e + T_{\text{proc}} + T_{\text{period}}) \\
&+[1 - (1 - p(r))^4](7T_d^e + T_{\text{period}} + T_{\text{proc}} + T_{\text{new}})\}
\end{aligned}
\tag{10.21}
$$

where:
T_{proc} is the time that the new AP and the MS take to build a new wireless link

Based on the comparison between the expectation of the communication delay and the communication cycle T, we can get the reward function at each communication cycle according to the quadric cost (Equation 10.10).

$$
r_k = \frac{1}{(x_k - \tilde{x}_k)^T Q(x_k - \tilde{x}_k) + u_k^T R u_k + \beta G_k}
\tag{10.22}
$$

$$
G_k = F\left(x_k^f - x_k^{fc}\right)^T \left(x_k^f - x_k^{fc}\right)
$$

where:
x_k^f is the actual state (position and velocity) of the front train
x_k^{fc} is the derived state of the front train from the MA from ZC by the current train
F is a counter that indicates the quantity relationship of the total time delay and the communication cycle, and it is defined as

$$
F = \begin{cases}
1, T_c <= T \\
i, 1 <= i <= \lceil \frac{T_c}{T} \rceil, T_d > T
\end{cases}
\tag{10.23}
$$

where T_c is the total time delay, and F decreases 1 at each communication cycle during the time delay.

Specifically, when analyzing the impact of the communication delay, both the uplink delay and the downlink delay should be considered. Hence, T_{fd}^e and T_{fh}^e are defined as the time delay (without any handoff) and the handoff latency of the uplink. Similarly, the performance parameters of the downlink are defined as T_{cd}^e

and T_{ch}^e. x_k is the desired state of the train, which includes the position information p_k and the velocity information v_k. The tracking error

$$x_k - \tilde{x}_k = \begin{bmatrix} p_k - \tilde{p}_k \\ v_k - \tilde{v}_k \end{bmatrix}$$

is determined by the uplink delay and the downlink delay. When ZC sends the MA to the current train, the downlink delay directly affects the transmission of the MA. However, MA is generated based on the information of the front train, which means that the transmission of the MA is indirectly affected by the uplink delay. Generally, at the first communication cycle, the front train sends its information to ZC and the current train gets the MA from ZC at the next communication cycle.

As the communication delay without handoffs is determined by the retransmissions of one data packet, whereas the handoff latency is related to the overall handoff procedure (seven frames), the communication delay (without handoff) and handoff latency are not considered at the same time when we study the impacts of time delay on the information gap. Then, we define $T_{ch}^e = 0$ or $T_{fh}^e = 0$, when there is no handoff. Similarly, $T_{cd}^e = 0$ or $T_{fd}^e = 0$, when the handoff procedure happens. As a result, we take $T^d = T_{ch}^e + T_{cd}^e$ as the downlink delay and $T^u = T_{fh}^e + T_{fd}^e$ as the uplink delay. When $(T^u/T) < 1$, the information of the front train can be sent to ZC in one communication period. When $(T^u/T) < 1$, the MA can be received by the current train in one communication period. As mentioned earlier, the MA is sent in a communication period after the information of the front train is received by ZC. Then, the tracking error $\Delta x_k = x_k - \tilde{x}_k$ is determined by one communication cycle. However, when $(T^u/T) >= 1$ or $(T^d/T) >= 1$, the scenario is complicated. We define $\lfloor T^u/T \rfloor = N_u$ or $\lfloor T^d/T \rfloor = N_d$, where $N_u \in N^*$ and $N_d \in N^*$. $H_k^d \in [1 \ N_d]$ and $H_k^u \in [1 \ N_u]$ are the indicators of the current communication cycles during the downlink delay and the uplink delay, respectively. Now we denote the acceleration of the front train and the current train as a_k^f and a_k^c, respectively, and the velocity as v_k^c and v_k^f. We can get the tracking error when the time delay is larger than the communication cycle as follows:

$$\Delta x_k = \begin{bmatrix} \Delta p_k \\ \Delta v_k \end{bmatrix} = \begin{bmatrix} p_k - \tilde{p}_k \\ v_k - \tilde{v}_k \end{bmatrix}$$

$$= \begin{bmatrix} \Delta p_{k-H_k^d} + \sum_{i=k-H_k^d}^{k} \left(\Delta v_i^c T + \frac{1}{2} \Delta a_i^c T^2 \right) \\ \Delta v_{k-H_k^d} + \sum_{i=k-H_k^d}^{k} \left(\Delta a_i^c T \right) \end{bmatrix} \quad (10.24)$$

where:

Δa_k^c is the difference of the actual acceleration and the desired acceleration, where the difference is determined by

$$\Delta x_k^f = x_k^f - x_k^{fc} = \begin{bmatrix} \Delta p_k^f \\ \Delta v_k^f \end{bmatrix}$$

where:

Δp_k^f is the distance that the front train runs during the time delay
Δv_k^f is the velocity variant range of the front train during the time delay

The two variables are determined by H_k^u and H_k^d.

$$HH = H_k^u + H_k^d + 1 - \delta$$

$$\Delta p_k^f = \sum_{j=k-HH}^{k} \left(v_j^f T + \frac{1}{2} a_j^f T^2 \right) \qquad (10.25)$$

$$\Delta v_k^f = \sum_{j=k-HH}^{k} \left(a_j^f T \right)$$

where:

δ is the overlapping number of communication cycles between the uplink delay and the downlink delay

10.5 Simulation Results and Discussions

In the section, we present simulation results to show the performance of the proposed cognitive control approach to CBTC systems. We first present the details of simulations, including the train dynamics, ATO, and wireless channel. Next, the simulation results are discussed.

10.5.1 Parameters of Train Dynamics

According to train dynamics, the train state space equation can be written as

$$p_{k+1} = p_k + v_k^c T + \frac{1}{2} \frac{u_k}{M} T^2 - \frac{1}{2} \frac{w_k}{M} T^2$$

$$v_{k+1}^c = v_k^c + \frac{u_k}{M} T - \frac{w_k}{M} T \qquad (10.26)$$

where:

M is the train mass
w_k is the resistance including slope resistance, curve resistance, and wind resistance
u_k is the control command generated from ATO

As a result,

$$A = \begin{bmatrix} 1 & T \\ 0 & 1 \end{bmatrix}, \quad B = \begin{bmatrix} \dfrac{1}{2M}T^2 \\ \dfrac{1}{M}T \end{bmatrix}, \quad C = \begin{bmatrix} -\dfrac{1}{2M}T^2 \\ -\dfrac{1}{M}T \end{bmatrix}$$

according to Equation 10.7.

10.5.2 Parameters of the ATO

According to Figure 10.5, formulae (10.7 and 10.8), ATO plays an important role in the train operation, which gives the control command according to the state of two trains y_k, which is defined as $\left[\tilde{p}_k^{fc}, p_k, \tilde{v}_k^{fc}, v_k \right]^T$, where $\tilde{p}_k^{fc}$ and $\tilde{v}_k^{fc}$ are the position and the velocity of the front train derived from the MA received by the current train, respectively, and p_k and v_k are the actual position and velocity of the current train, respectively.

Based on the optimal train running profile in [28], we divide the train operation process into three steps: acceleration, coasting, and braking. As a result, there are two switching points sp_1 and sp_2. Then, the ATO model of Equation 10.8 can be converted as follows:

$$x_{k+1}^c = B^c y_k - T_c$$

$$u_k = C^c f(x_k^c) \tag{10.27}$$

where:

$T_c = \left[L, sp_1, sp_2, v_{\max} \right]^T$ contains the constant parameters related to the subway line and the design standard

L is the safe distance, which should be kept between adjacent trains

$v_{\max}$ is the maximum limited velocity of the subway line

$$B^c = \begin{bmatrix} 1 & -1 & 0 & 0 \\ 0 & 1 & 0 & 0 \\ 0 & 1 & 0 & 0 \\ 0 & 0 & 0 & 1 \end{bmatrix} \tag{10.28}$$

As the train operation profile is partitioned into three parts, there are different control schemes corresponding to each step. Generally, at the acceleration step, the acceleration is a constant positive number to make the train reach the maximum speed as soon as possible; at the coasting step, there is no power acting on the train, which moves through inertia; at the last step, the train should brake with a constant deceleration until it reaches the safe stopping point. The control method can bring the minimum energy consumption and the maximum efficiency. However, considering the system with two or more trains, due to the safe distance and the limitation of the optimal running profile, if the distance between the adjacent trains is small because of communication delay, the back train has to brake in order to keep safe with the deceleration calculated by the ATO. As a result, we define C^c as $[A_c, 0, A_b]$, where A_c is the constant acceleration and A_b is the braking deceleration. After the ATO processes the state vector y_k, the x_k^c is obtained to determine the relationship between the state of the current train and the state of the front train. As the optimal train operation profile is piecewise, we employ logical operations on the elements of x_k^c and $f(x_k^c)$ as follows:

$$f(x_k^c) = \begin{bmatrix} x_k^1 \wedge [x_k^2 \vee (x_k^2 \veebar x_k^3 \wedge x_k^1 \wedge x_k^4)] \\ x_k^1 \wedge x_k^2 \wedge (\neg x_k^3) \wedge \neg (x_k^2 \veebar x_k^3 \wedge x_k^1 \wedge x_k^4) \\ x_k^2 \wedge x_k^3 \vee \neg (x_k^1) \end{bmatrix} \quad (10.29)$$

where:
$x_k^c = [x_k^1, x_k^2, x_k^3, x_k^4]$
$\wedge$ is the logical AND
$\vee$ is the logical OR
$\veebar$ is the logical XOR
$\neg$ is the logical NOT

As a result, according to the relationship between the states of two trains and some constant parameters of the optimal train operation profile, ATO can determine the acceleration of the train at each communication cycle.

10.5.3 Parameters of the Wireless Channel

According to the measurement results in Beijing Yizhuang Line, we derive the state transition probability for each interval. The FSMC model is built with four states and the 5 m distance interval. Figure 10.7 shows the measurement scenario.

According to the measurement results, one of the channel state transition matrices is

$$P = \begin{bmatrix} 0.91 & 0.08 & 0 & 0 \\ 0.041 & 0.86 & 0.09 & 0 \\ 0 & 0.024 & 0.85 & 0.11 \\ 0 & 0 & 0.023 & 0.97 \end{bmatrix}$$

which shows the channel characteristics at the location 35–40 m.

10.5.4 Simulation Results and Discussions

In this section, simulation results are presented and discussed. First of all, we present the train control performance improvement. Next, the handoff performance is discussed. In addition, we show that the proposed cognitive control approach can increase the reliability of train–ground communication, which is also an important parameter for CBTC systems.

Figure 10.7 (a) Tunnel where we performed the measurements. (b) Shark-fin antenna located on the measurement vehicle. (c) Yagi antenna. (d) AP set on the wall.

We implement the simulations using MATLAB. As mentioned earlier, we get the channel state probability through real field measurements. In our simulation scenarios, there are two stations and the distance is 2256 m, which is the real value of the distance between Tongji Nan and Jinghai stations in Beijing Subway Yizhuang Line, and the regulated trip time is 150 s. According to the deployment of wayside APs, the length of interval between two adjacent APs is 400 m. As a result, there are six APs between these two stations. In the simulations, there are two trains. The headway is first set to 15 s, which means that the second train departs from the starting station 15 s after the first train leaves. The headway of 90 s is also considered in the simulations. The parameters related to the dynamic model and the wireless channel model are illustrated in Table 10.1.

Table 10.1 Availability under Different Policies

Policy	Availability (A_{av})	Unavailability ($1 - A_{av}$)
Cognitive control	0.9978	2.2×10^{-3}
SMDP	0.9413	5.87×10^{-2}
Greedy	0.8833	1.167×10^{-1}

There are three policies in our simulations for comparisons: the proposed cognitive control policy, the semi-Markov decision process (SMDP) policy, and the greedy policy. Based on the Markov property of state transition process, it is possible to model the problem considered in this chapter as an SMDP [29] and derive the SMDP policy. In the greedy policy, if there is one AP whose signal strength is higher than the current associated AP, the MS switches to the AP with higher signal strength. In other words, the greedy policy always makes decisions based on the immediate reward, not the long-term reward.

First of all, we compare the cost function under different policies. As shown in Figure 10.8, the x-axis represents the index of the communication cycle and the y-axis is the cost in each communication cycle. Under the greedy and SMDP policies, the cost increases sharply in some communication cycles, which means that the information gap becomes larger due to the long handoff latency. Obviously, the SMDP policy can bring better performance and less cost with less peaks compared with the greedy policy. However, the cost under the proposed cognitive control policy is a smooth curve, which means that no long handoff latency happens. Figure 10.8 indicates that the cognitive control can help the MS to make the optimal handoff decision through minimizing the information gap, which can decrease the cost of train control including the tracking errors and energy consumption.

The travel trajectories of these two trains under different policies are shown in Figures 10.9 through 10.11, where the x-axis is the position of trains and the y-axis is the corresponding velocity of trains. Under different policies, the current (back) train follows different travel curves. When the greedy policy and the SMDP policy are used, the train will be off the preset running profile sometimes due to the large information gap. The handoff latency enlarges the information gap, and the current train has to slow down in order to keep the safe distance. Next, when the latest MA is received by the train and the information gap is eliminated, the current train has to speed up to reach the optimal running profile. As there are frequent accelerations and decelerations, it can cause much more energy consumption. By contrast, as shown in Figure 10.11, the current train with the cognitive control policy can be very close to the optimal running profile, which means improved passenger comfort and energy saving in the proposed scheme.

We also consider the case when the headway is 90s, which is the standard headway used in Beijing Yizhuang Line. From Figures 10.12 through 10.14,

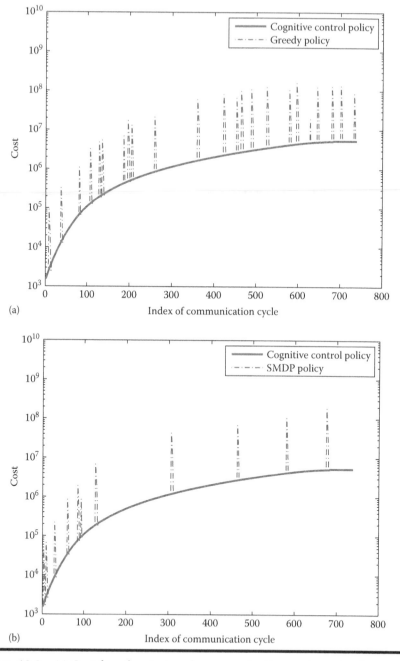

(a)

(b)

Figure 10.8 (a) Cost function *J* at each communication cycle under the greedy policy and the proposed cognitive control policy. (b) The cost function *J* at each communication cycle under the SMDP policy, and the proposed cognitive control policy.

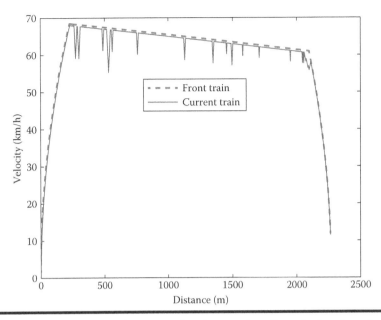

Figure 10.9 **Train travel trajectory under the greedy policy (the headway is 15 s).**

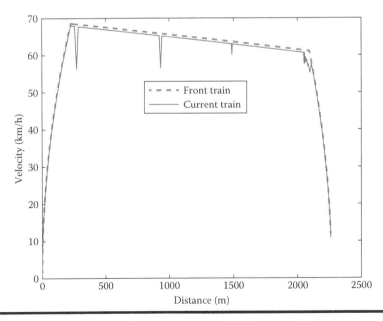

Figure 10.10 **Train travel trajectory under the SMDP policy (the headway is 15 s).**

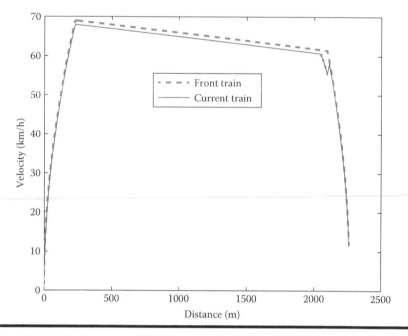

Figure 10.11 **Train travel trajectory under the proposed cognitive control policy (the headway is 15 s).**

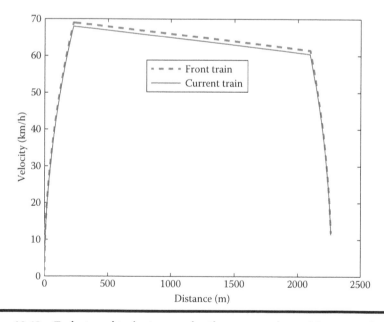

Figure 10.12 **Train travel trajectory under the proposed cognitive control policy (the headway is 90 s).**

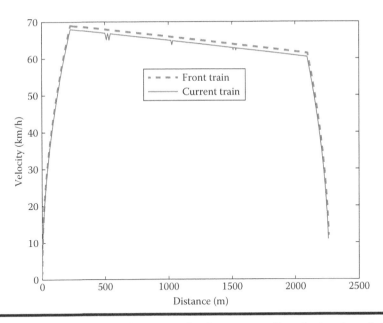

Figure 10.13   Train travel trajectory under the SMDP policy (the headway is 90 s).

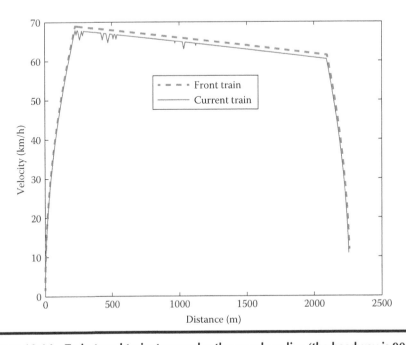

Figure 10.14   Train travel trajectory under the greedy policy (the headway is 90 s).

we can see that the effects caused on the communication latency are less severe when the headway is 90 s, compared to the case when the headway is 15 s. Nevertheless, the proposed cognitive control approach can improve the performance of train operation in both cases, compared to the existing SMDP policy and greedy policy.

The handoff performance is presented in Figure 10.15, where the *x*-axis is the location of the train and the *y*-axis is the value of handoff latency. It is obvious that the proposed cognitive control policy can bring less handoff delay, which is less than the communication cycle 0.2 s. By contrast, the handoff performance under the greedy policy and the SMDP is worse, where the handoff delay could be more than 1 s. In addition, there are less ping-pong handoff happening under the cognitive control policy. Therefore, the handoff performance is improved due to the application of the cognitive control approach.

As CBTC systems are safety critical, the reliability of train–ground communication is an important performance parameter. Hence, according to the handoff latency, we calculate the failure rate [30] of the train–ground communication subsystem. Failure rate describes the frequency with which an engineered system or component fails and is important in reliability engineering. Figure 10.16 shows the failure rate with time. The cognitive control approach can keep the failure rate less than 10^{-6}, which means that the reliability is largely increased. Considering the

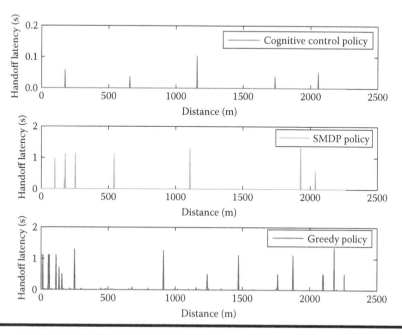

Figure 10.15 Handoff latency under different policies.

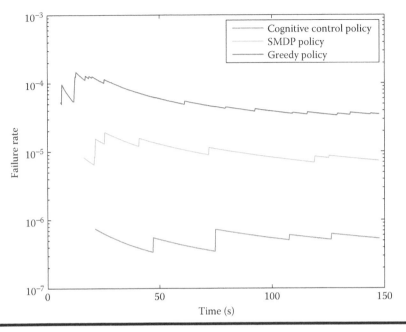

Figure 10.16   Train–ground failure rate under different policies.

frequency of handoff and the value of handoff latency, we can derive the availability of the train–ground communication subsystem as follows [31]:

$$A_{av} = \frac{MTTF}{MTTF + MTTR} \tag{10.30}$$

where:

MTTF (mean time to failure) denotes the mean time between adjacent handoffs

MTTR (mean time to repair) denotes the mean value of handoff latency

The availability of different policies is shown in Table 10.2.

The cognitive control approach can get the highest availability 0.9978. In other words, the unavailability under the cognitive control policy can be kept at an order of magnitude 10^{-3}, whereas it is 10^{-2} under the SMDP policy and 10^{-1} under the greedy policy. We can conclude that the application of cognitive control can get significantly better train control performance, improved handoff performance, and reliability of CBTC systems compared with other policies.

Figure 10.17 shows the learning procedure in the proposed cognitive control approach. Specifically, the difference between adjacent operation policies versus the steps of Q-learning is shown in Figure 10.17. At the beginning, the learned policies

Table 10.2   Parameters Used in the Simulations

Emergency brake deceleration	1.2 m/s^2
Service brake deceleration	0.8 m/s^2
Tracking acceleration	0.8 m/s^2
The response time of the train	0.4 s
The running resistance per unit mass	0.02 m/s^2
The limited line speed	80 km/h
aSifsTime	10 μs
aSlotTime	9 μs
anACKTime	1 μs
aDifsTime	aSifsTime+2aSlotTime
CW$_{min}$	16
CW$_{max}$	1024
aPacketLength	400 bytes
aPropTime	1 ms

at adjacent steps are far from the optimal one, and they are quite different. After about 180 steps, the difference between adjacent policies is zero, which means that the learned policy converges to the optimal one.

10.6 Conclusion

In this chapter, we presented a cognitive control approach to CBTC systems to improve the train control performance, considering both train–ground communication and train control. In the proposed cognitive control approach, we introduced information gap, which is defined as the difference between the derived state of the front train and the actual state of the front train in CBTC systems. Linear quadratic cost for the train control performance in CBTC systems was considered in the performance measure. In addition, information gap was formulated in the cost function of cognitive control to quantitatively describe the effects of train–ground communication on train control performance. Based on the cognitive control formulation, RL was used to get the optimal policy. Moreover, the wireless channel was modeled as FSMCs with multiple state transition probability matrices,

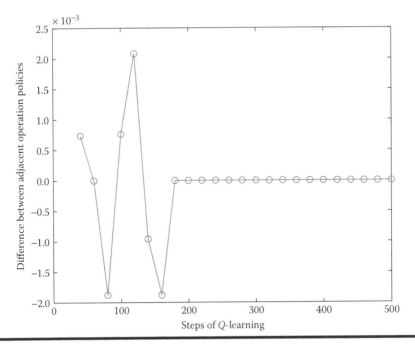

Figure 10.17   Performance of optimization versus steps.

which can bring much more accuracy than the model with only one state transition probability matrix. Simulation results were presented to show that the cognitive control approach can significantly improve the performance of train control compared with other policies.

References

1. R.D. Pascoe and T.N. Eichorn. What is communication-based train control? *IEEE Veh. Tech. Mag.*, 4(4):16–21, 2009.
2. IEEE. Standard for communications-based train control (CBTC) performance and functional requirements. *IEEE Std 1474.1-2004 (Revision of IEEE Std 1474.1-1999)*, pages 0_1–45, New York, 2004.
3. M. Heddebaut. Leaky waveguide for train-to-wayside communication-based train control. *IEEE Trans. Veh. Tech.*, 58(3):1068–1076, 2009.
4. L. Zhu, F.R. Yu, B. Ning, and T. Tang. Cross-layer handoff design in MIMO-enabled WLANs for communication-based train control (CBTC) systems. *IEEE J. Sel. Areas Commun.*, 30(4):719–728, 2012.
5. L. Zhu, F.R. Yu, B. Ning, and T. Tang. Handoff performance improvements in MIMO-enabled communication-based train control systems. *IEEE Trans. Intell. Transp. Sys.*, 13(2):582–593, 2012.

6. J. Huang, J. Ma, and Z. Zhong. Research on handover of GSM-R network under high-speed scenarios. *Railway Commun. Signals*, 42:51–53, 2006.

7. C. Yang, L. Lu, C. Di, and X. Fang. An on-vehicle dual-antenna handover scheme for high-speed railway distributed antenna system. In *Proc. IWCMC*, Chengdu, China, September 2010.

8. L. Zhu, F.R. Yu, B. Ning, and T. Tang. Cross-layer handoff design in MIMO-enabled WLANs for communication-based train control (CBTC) systems. *IEEE J. Sel. Areas Commun.*, 30(4):719–728, 2012.

9. K. Rahn, C. Bode, and T. Albrecht. Energy-efficient driving in the context of a com-munications-based train control system (CBTC). In *Proc. IEEE Int'l Conf. Intelligent Rail Transportation*, Beijing, China, August 2013.

10. J. Tang and X. Zhang. Cross-layer resource allocation over wireless relay networks for quality of service provisioning. *IEEE J. Sel. Areas Commun.*, 25(4):645–657, 2007.

11. F.R. Yu, B. Sun, V. Krishnamurthy, and S. Ali. Application layer QoS optimization for multimedia transmission over cognitive radio networks. *ACM/Springer Wireless Networks*, 17(2):371–383, 2011.

12. B. Bu, F.R. Yu, and T. Tang. Performance improved methods for communication-based train control systems with random packet drops, *IEEE Trans. Intell. Trans. Syst.*, 15(2):1179–1192, 2014.

13. S. Haykin. *Cognitive Dynamic Systems: Perception-Action Cycle, Radar and Radio*. Cambridge University Press, Cambridge, UK, 2012.

14. S. Haykin, M. Fatemi, P. Setoodeh, and Y. Xue. Cognitive control. *Proc. IEEE*, 100(12):3156–3169, 2012.

15. R.B. Mars, J. Sallet, M.F.S. Rushworth, and N. Yeung. *Neural Basis of Motivational and Cognitive Control*. MIT Press, Cambridge, MA, 2012.

16. S. Sastry and M. Bodson. *Adaptive Control: Stability, Convergence and Robustness*. Dover Publications, Upper Saddle River, NJ, 2011.

17. T. Hrycej. *Neurocontrol: Towards an Industrial Control Methodology*. Wiley, New York, 1997.

18. S. Haykin. Cognitive dynamic systems. *Proc. IEEE*, 94(11):1910–1911, 2006.

19. F. Yu, V.W.S. Wong, and V. Leung. Efficient QoS provisioning for adaptive multime-dia in mobile communication networks by reinforcement learning. *Mobile Net. Appl.*, 11(1):101–110, 2006.

20. T.M. Mitchell. *Machine Learning*. McGraw Hill, New York, 1997.

21. S. Lin, Z. Zhong, L. Cai, and Y. Luo. Finite state Markov modelling for high speed rail-way wireless communication channel. In *Proc. IEEE Globecom'12*, Anaheim, CA, 2012.

22. F. Babich, G. Lombardi, and E. Valentinuzzi. Variable order Markov modeling for LEO mobile satellite channels. *Electron. Lett.*, 35(8):621–623, 1999.

23. H.S. Wang and N. Moayeri. Finite-state Markov channel—A useful model for radio communication channels. *IEEE Trans. Veh. Tech.*, 44(1):163–171, 1995.

24. P. Sadeghi, R. Kennedy, P. Rapajic, and R. Shams. Finite-state Markov modeling of fading channels—A survey of principles and applications. *IEEE Signal Proc. Mag.*, 25(5):57–80, 2008.

25. L. Zheng and D.N.C. Tse. Diversity and multiplexing: A fundamental tradeoff in multiple-antenna channels. *IEEE Trans. Inform. Theory*, 49(5):1073–1096, 2003.

26. D. Gesbert, M. Shafi, D.S. Shiu, P.J. Smith, and A. Naguib. From theory to prac-tice: An overview of MIMO space-time coded wireless systems. *IEEE J. Sel. Areas Commun.*, 21(3):281–302, 2003.

27. IEEE. IEEE standard for information technology–telecommunications and information exchange between systems local and metropolitan area networks–specific requirements part 11: Wireless LAN medium access control (MAC) and physical layer (PHY) specifications. *IEEE Std 802.11-2012 (Revision of IEEE Std 802.11-2007)*, pages 1–2793, New York, 2012.

28. S. Su, X. Li, T. Tang, and Z. Gao. A subway train timetable optimization approach based on energy-efficient operation strategy. *IEEE Trans. Intell. Trans. Syst.*, 14(2):883–893, 2013.

29. M.L. Puterman. *Markov Decision Processes: Discrete Stochastic Dynamic Programming*. John Wiley & Sons, New York, 1994.

30. T.H. Xu, T. Tang, C.H. Gao, and B.G. Cai. Dependability analysis of the data communication system in train control system. *Science in China Series E: Technological Sciences*, 52(9):2605–2618, 2009.

31. M. Rausand and A. Høyland. *System Reliability Theory: Models, Statistical Methods, and Applications*. John Wiley & Sons, New York, 2004.

Index

Note: Locator followed by '*f*' and '*t*' denotes figure and table in the text

3rd Generation Partnership Project (3GPP), 151

A

Access points (APs), 97
  antenna, 157
  candidate, 155, 159
  coverage area, 188–190, 190*t*
    communication, 224
    overlapping, 191, 191*f*
  delay difference between adjacent, 160, 160*f*
  output power of, 83
Action space, 131
Active handover scheme, 187
Active link
  failure of, 97
  between train and ground, 97–98
Active scanning mode, 152
Adaptive modulation and coding (AMC), 167, 172
Akaike information criterion (AIC), 82
  general case, 74
  second-order, 88
  selecting candidate distribution, 75, 75*f*, 88, 89*f*
Alamouti code, 156, 158
Alternating current field measurement (ACFM), 57
Antenna
  AP, 157
  gain of, 83
  head directional, 97
  leaky wave, 82
  receiving, 68, 90
  Shark-fin, 68, 70*f*, 234*f*
  transmitting, 67–68

tunnel section and deployment of, 68, 69*f*
  two directional, 97–98
  Yagi, 67–68, 70*f*, 234*f*
Array signal processing, 157
Asynchronous dynamical systems (ADSs), 198
Automated visual inspection, 51–52
Automatic repeat request (ARQ) scheme, 128, 186
Automatic TP (ATP), 45
Automatic train protection (ATO) system, 2, 18, 119, 217, 221
  onboard, 7
  parameters of, 232–233
  tests, 35
  in train operation, 232
  wayside, 6
Automatic train protection (ATP) system, 2, 18, 217, 221
  onboard, 6
  subsystem, 119, 123
  wayside, 6
Automatic train supervision (ATS) system, 2, 6, 18, 21
  advanced, 35
  subsystem, 120
  tests, 35
Azimuthal scanning, 54

B

Backhaul networks, 121–122, 127
Backup link, 97–98, 98*f*, 99*f*, 100*f*
Base stations, CBTC systems, 130
  active, 126, 128
  cluster of, 117
  implementation of dynamic, 117

Beam steering, 54
Beijing Subway Changping Line, 68, 70*f*, 76,
     186, 201
Beijing Subway Yizhuang Line
   leaky waveguide applied in, 82, 83*f*
   measurement results in, 233, 234*f*
   packet drop rate on, 201
   retransmission times of, 186
Binary phase shift keying (BPSK), 158
Bit error rate (BER), 158
   data transmission rate and, 126–128
   probability, 224
   real system, 128
   target, 164
Bolt hole cracks, 49
Break-before-make handoff scheme, 167–168
Brownfield project, 20–21, 28, 38

C

Call admission control (CAC), 94
Candidate AP, 155, 159
Candidate models, 75, 88
Carborne controller (CC), 17, 21, 25
   installation verification, 29
   tests, 28
Carrier sense multiple access/collision avoidance
     (CSMA/CA), 159, 226
   handoff signaling packet in, 161–162
   random backoff scheme in, 186
Catastrophic accidents, 44
Catastrophic rail failure, 50
CBTC WLAN system, 151, 159
Channel 1, 83
Channel matrix, 127, 157
Channel models, 85
Channel scanning scheme, 150
Cisco 3200 routers, 67
Cisco WLAN devices, 83
Closed-loop control systems, design of, 202–203
Cluster transition probability, 133
Cognitive control approach, 218–220
   actions, 219
   advantages, 216
   to CBTC systems, 217–218
   cost function, 218–219
   failure rate, 240
   formulation of, 221–231
        channel model in MIMO-enabled
            WLANs, 223–225
        Q-learning in, 225–231
        train control model, 221–223

overview of, 216–218
parameters
   of ATO, 232–233
   in simulations, 241, 242*t*
   of train dynamics, 231–232
   of wireless channel, 233
policy, 235, 236*f*, 238*f*
procedure of, 217, 217*f*
RL model in, 225, 225*f*
simulation results and discussions, 233–242
structure of, 216, 216*f*, 221, 222*f*
sufficient information in, 217
Cognitive radar/radio systems, 216
Commercial off-the-shelf (COTS)
     equipment, 214
Communication channel model, 125
Communication delay, 226
   analyzing impact of, 229
   tracking error, 230
   without handoffs, 230
Communication latency, 128–129
   on CBTC efficiency, 119–120
   handoff, 120
   long, 153
   short, 153
   in train–ground communication systems,
        119, 129
   in WLANs, 161–162
Communications-based train control (CBTC)
     system, 1
   APs in, 154
   architecture, 21
   base stations, 130
   capacity, 45–46
   challenges of, 7–8
   channel measurements
        equipment, 67–68, 68*f*
        objective of, 67
        preparation, 67
        scenario, 68, 69*f*
   characteristics of, 66
   communication latency on
        with CoMP, 126–129
        impacts of, 119–121
   CoMP, 121–122, 122*f*
        adoption of, 117–118
        control performance optimization,
             129–139
   cons of, 19
   features and architecture of, 5–7, 6*f*
        closed-loop control, 5
        GoA, 5

onboard equipment, 6–7
train–ground radio communication
  subsystem, 7
wayside equipment, 5–6
goal of, 4
handoff procedure
  communication latency, 152–154
  features of, 152–154
  handoff latency of 802.11, 152–154
IEEE 1474 standard defined, 2, 45
impacts of wireless communications on,
  119, 120*f*
information gap of, 217
MAL, 4, 4*f*
measurement campaign, 83–84, 84*f*
minimum headway of, 178
objective of, 1–2
packet-based transmission, 124
packet drops in. *See* Packet drops in CBTC
projects
  installation on new line, 19–20
  migrating existing line, 20–21
projects of, 8, 13
pros of, 18–19
radio channel with leaky waveguide
  modeling of, 84–91
  overview, 82–83
radio waves of, 82
required data rate in, 158
security, 8
technologies of, 178–179
test duration, 38
testing principles, 21–22
trains' control in. *See* Trains' control in
  CBTC systems
WLAN-based, 94
CoMP-based greedy scheme w/o updated
  trajectory, 140
CoMP-based scheme w/o updated trajectory, 140
Constraints, 134
Contention window (CW) timer, 155
Continuous bidirectional wireless
  communications, 96
Continuous-time Markov chain (CTMC)
  model, 94
  and DCS, 96–97
    with basic configuration, 100, 100*f*
    first proposed, 100, 100*f*, 102*f*
    numerical results and discussions, 110–112
      availability improvement, 111–112
      model soundness, 111
      system parameters, 110–111, 110*t*

theory, 99
vs. DSPN for redundancy configurations,
  111, 111*f*
Conventional ultrasonic rail inspection, 53
Coordinated multipoint (CoMP) method, 117
  communication latency in CBTC systems
    with, 126–129
    coordinated multipoint transmission and
      reception, 126
    data transmission rate and BER,
      126–128
  control performance optimization in
    CBTC, 129–139
    handoff decision model, 129–134
    optimal guidance trajectory calculation,
      137–139, 137*f*
    SMDP-based CoMP cluster selection,
      129–134
  enabled CBTC, 117
  optimal policy for, 139
  simulation results and discussions, 139–145
    handoff performance improvement,
      143–145, 145*f*
    parameters, 139, 139*t*
    train control performance improvement,
      140–143
Corrugation, 49

D

Data communication system (DCS), 21
  tests, 30–32
    radio, 31–32
    wayside network, 30–31
  WLAN-based, 94
    availability analysis, 99–102
    basic configuration of, 96, 96*f*
    behavior with DSPNs, 103–110
    first proposed, 97, 98*f*
    overview of CBTC and, 96–97
    with redundancy, 97–99
    second proposed, 98, 99*f*
    unavailability of, 111, 112*f*
    wireless portions of, 96
Data rate, 127
Decision epochs, 130
Deterministic and stochastic Petri net (DSPN), 94
  DCS behavior with, 103–110
    basic configuration, 104, 104*f*
    formulation, 104–105
    introduction to, 103–104
    model solutions, 105–110

Deterministic and stochastic Petri net (DSPN)
(*Continued*)
redundancy and backup link, 105, 105*f*
redundancy and no backup link, 105, 105*f*
EMC for, 106
marking process of, 105–106
motivations of selecting, 103
vs. CTMC for redundancy configurations,
111, 111*f*
Distance-to-go technology, 3
Distortion function, 73
Distributed interframe space (DIFS), 155
Diversity gain, 224
Downlink delay, 229–230
Downlink transmissions, 183–184
Dynamic model
parameters related to, 234, 235*t*
train, 132, 221
Dynamic PICO test, 28

E

Eddy current rail inspection, 56
Effective isotropic radiated power (EIRP), 189
Elastic fastening systems, 49, 50*f*
Electromagnetic acoustic transducers
(EMATs), 57
ElectroMagnetic Compatibility (EMC) tests, 23
Electronic focusing, 54
Electronic scanning, 54
Embedded Markov chain (EMC)
for DSPN, 106
steady-state probability for, 109
Empirical cumulative distribution functions
(CDFs), 88, 89*f*
Environmental tests
abrasive conditions, 24
climatic conditions, 23
EMC test, 23
mechanical conditions, 24
Equiprobable partition method, 72
Equivalent magnetic dipole method, 82
simulation results of, 86, 87*f*
Equivalent NCS, 183–184, 183*f*
Error-free period, 164–166
of communication system, 165
of traditional handoff schemes, 165,
169–170
Error-free transmission period, 166
European train control systems (ETCSs),
46, 95
Existing scheme, 140

F

Factory acceptance test (FAT), 26, 39
Factory tests, 24–27
factory setup, 25
goals, 25
types of
CBTC supplier internal, 26
challenges of, 27
FAT, 26
product, 26
tests to be performed, 26–27
Fastening parts, rail
comparison of NDT methods, 57, 61*t*
defects in, 49
elastic, 49, 50*f*
infrastructure, 46–50
critical place on rail network, 49–50
fastening parts overview, 49
maintenance, 46
rail overview, 46–49
superstructure subsection, 46, 47*f*
inspection, 50–57
ACFM, 57
advantages and disadvantages of, 57,
58*t*–60*t*
eddy current, 56
liquid penetrant, 52
magnetic flux leakage, 55–56
manual and automated visual, 51–52
ultrasonic, 52–55, 53*f*
using EMATs, 57
rigid, 49, 50*f*
Field tests, constraints on, 38–39
Finite-state Markov channel (FSMC) model,
69–76, 125
accuracy of, 76
distribution of SNR, 74–76
features of, 66–67
MSE between experimental data and, 77, 78*f*
range of received SNR, 125
real field measurement, 76–79
received SNR, 223
simulation results, 77, 78*f*
SNR-level thresholds of, 72–74, 76*t*, 77*t*
state transition probabilities of, 76, 77*t*
First Article Configuration Inspection (FACI),
24
Fixed block systems, 3
Fixed-block track circuit-based train control
systems, 137
Fixed sensors on rail, 55, 55*f*

Frame error rate (FER), 151, 226
  at 6.5 Mbps, 170, 170*f*
  at 13 Mbps, 170, 171*f*
  at 26 Mbps, 170, 171*f*
  at 65 Mbps, 170, 172*f*
  of handoff signaling, 166–167
    with data rates, 170–173
    inter-site distance to meet HO, 172, 173*f*
    of MAHO, 172
    train speeds, 186, 186*f*
  of transmission schemes, 158, 159*f*
Functional tests, 33–35

G

Generalized SPNs (GSPNs), 103–104
Ghost mode, 36
Global System for Mobile Communications-
    Railway (GSM-R), 94, 153
Grades of Automation (GoA), 5
Greedy policy, 235, 236*f*
  handoff performance, 240
  train travel trajectory under, 235, 237*f*, 239*f*
Greenfield project, 19–20
Guided wave testing method, 54–55

H

Half power beam width (HPBW), 67
Handoff algorithm, 154, 214
Handoff schemes
  action, 226
  communication latency, 120
  decision model, 129–134
    actions, 130–131
    constraints, 134
    decision epochs, 130
    reward function, 131–132
    solutions to, 134–137
    states, 131
    state transition probability, 132–134
  decision problem, 117
  error-free periods of traditional, 169–170
  handoff latency, 228
    of 802.11, 152–154
    analysis, 167–169
    calculation, 162
    expectation of, 229
    failure rate, 240
    under policies, 240, 240*f*
  location-based, 150–151, 154
  performance analysis of, 162–167

error-free period, 164–166
FER of handoff signaling, 166–167
impacts on ongoing data sessions, 167
optimal handoff location, 163–164
simulation results and discussions,
    167–173, 168*t*
wireless channel model, 162–163
performance improvement, 143–145
policy, 122, 137, 144*f*
procedure
  basic WLAN, 228, 228*f*
  normal, 162
  stages, 150
  steps, 228
  time interval between, 169, 169*f*
  transmitted frames, 228
proposed, 155, 156*f*
signaling
  BER of, 166–167
  with data rates, 170–173
  target FER of, 172
  triggering location, 160
Handover algorithm, 116
Handover scheme, packet drops due to
  AP's coverage area, 188–190
  overlapping coverage area, 191–193, 191*f*
  process, 187
  rate of packet drops, 191–193, 193*t*, 201*t*
  time, 187, 192*f*
Head directional antenna, 97
Host machine, test on, 25

I

IEEE 802.3 standard, 96
IEEE 802.11g WLANs, 226
IEEE 802.11 standard, 150
Infinitesimal generator matrix, 100–101, 108
Information gap, 217
Inspection. *See* Rail and fastening parts,
    inspection
Integrated design approach, 150
Integration tests, 26
  external to CBTC, 33
  internal to CBTC, 32–33
  on-board
    mechanical and electrical tests, 28
    rolling stock characterization tests,
        27–28
    static and dynamic PICO test, 28
Interference cancelation algorithm, 155, 158
Interlocking, wayside ATP, 6

International Council on Systems Engineering
(Book), 24
Intrusion detection systems (IDSs), 8

K

Kernel, DSPN, 106–107
Kolmogorov differential equation, 99
Kronecker function, 108

L

Laser ultrasonic rail inspection, 53–54
Leaky coaxial cable, 214
Leaky wave antenna, 82
Leaky waveguide
    in Beijing Subway Yizhuang Line, 82, 83*f*
    CBTC radio channel with
        modeling of, 84–91
        overview, 82–83
    length of, 83, 86–87
    transmission loss of, 86–87
Limit of movement authority (LMA), 5–6
Linear model of large-scale fading, 86
Linear quadratic cost function, 137
Line-of-sight (LOS), 160
Link state, 99
Liquid penetrant rail inspection, 52
Lloyd–Max algorithm, 73
Lloyd–Max technique, 72, 224
Localization tests, 32
Location-based handoff scheme, 150–151, 154
Log-normal distribution, 88
    expression of, 90
Long-range ultrasonics, 54–55
Long-Term Evolution (LTE), 151
Long-term evolution-advanced (LTE-A)
    systems, 117
Long view and minimized distance (Lv_d)
    scheme, 202, 204, 206*f*, 207*f*
Long view and minimized force (Lv_f) scheme,
    202, 204, 206*f*, 208*f*

M

Magnetic dipole antenna array, 86
Magnetic flux leakage rail inspection, 55–56
Manual rail inspections, 51–52, 51*f*
Markov decision process (MDP), 130
Markov property of state transition process, 130
Markov regenerative process (MRGP), 105
    steady-state probabilities of, 110

Markov renewal sequence, 106
MATLAB, 86, 234
    LMI toolbox of, 204
    MaxChannelTime, timer of, 160
    Maximum likelihood estimator (MLE)
        method, 90
Maximum ratio combining (MRC), 128
Mean square error (MSE), 77, 78*f*
Mean time between failures (MTBF), 36–37
Mean time between functional failures
    (MTBFF), 36–37
Mean time to repair (MTTR), 37
Mechanical and electrical tests, 28
Media access control (MAC), 161, 186, 226
MIMO-assisted handoff (MAHO) scheme, 151,
    154–162
    FER of, 172
    latency performance improvements of,
        167, 168*f*
    MIMO transmission in handoff procedure,
        155–161
    physical layer processing, 155–158
    synchronization in downlink, 159–161
Mobile station (MS), 67, 83, 221
Mobile terminal (MT), 116, 144
Modulation and coding scheme (MCS),
    158, 167
Monitor mode, 36
Movement authority (MA), 119, 217–218
Movement authority limit (MAL), 21
Multiple-input and multiple-output (MIMO),
    116, 151
    multiplexing packets, 155
    transmission in handoff procedure, 155–161
    physical layer processing, 155–158
    synchronization in downlink, 159–161
    transmission mode, 155, 163
    types of gains, 224
Multiple radio transceivers, 151
Multistate Markov model, 66

N

Nakagami distribution, 75
NDT techniques. *See also* Rail and fastening
    parts
    advantages and disadvantages of, 57,
        58*t*–60*t*
    comparison, 57, 61*t*
    defined, 51
Networked control system (NCS), 179
    equivalent, 183–184, 183*f*

Network management system (NMS), 30
Newton's Second Law, 221
Nondestructive testing (NDT) inspection, 45
Novel control schemes, 199–201
Novel handover scheme, 116

O

Odometer system, 32
Okumura–Hata model, 85
Onboard ATO, 7
Onboard ATP, 6
OnBoard Controller Unit (OBCU), 21
On-board integration tests
  mechanical and electrical tests, 28
  rolling stock characterization tests, 27–28
  static and dynamic PICO test, 28
Online value iteration algorithm, 135–137
On-site tests, 30–37
  ATO tests, 35
  ATS tests, 35
  DCS tests, 30–32
    radio tests, 31–32
    wayside network tests, 30–31
  functional tests, 33–35
  integration tests
    external to CBTC, 33
    internal to CBTC, 32–33
  localization tests, 32
  maintainability demonstration, 37
  PICO, 30
  reliability and availability demonstration,
    36–37
  shadow mode tests, 36
  site acceptance tests, 35–36
Operations Control Center (OCC), 21
Optimal guidance trajectory calculation,
  137–139, 137*f*
Optimal handoff location, 163–164
Optimal policy, 220
Optimal train running profile, 232–233
Original equipment manufacturer (OEM), 19
Orthogonal frequency division multiplexing
  (OFDM), 158, 161
Orthogonal STBC, 157

P

Packet drops in CBTC
  communication procedures with,
    180–181, 181*f*
  field test and simulation results, 201–208

closed-loop control systems, 202–203
  impacted by packet drops, 203–208
  on packet drop rate, 201, 201*t*
  simulation parameters, 202, 202*t*
in train–ground communications,
    186–193
  due to handover, 187–193
  due to random transmission errors,
    186–187
trains' control in CBTC with, 193–201
  currently used control scheme,
    193–194
  on performances of, 198–199
  on stability of, 197–198
  states estimation under, 194–197
in uplink transmissions, 182, 184
Packet loss, 167
  factors cause, 227
  retransmission of data packets, 226–227
  by transmission error, 161
Packet loss rate (PLR), 226
Passenger information system (PIS), 170
Passive scanning mode, 152
Path loss model, 66, 82, 162, 164
Petri net, 103
Phased array ultrasonic rail inspection, 54
Physical layer processing, 155–158
Ping-pong effect, 122
Ping-pong handoff, 143–144, 240
Positive train control (PTC), 45
  compliant, 18
Post Installation Check Out (PICO) tests, 30
  of network equipment, 31
  static and dynamic, 28
Product factory tests, 26

Q

Q-learning algorithm, 220, 225
Q-learning in cognitive control approach
  linear quadratic cost, 226
  reward function, 226–231
  system states and actions, 225–226
Quality-of-service (QoS)
  targets, 153
  trains' control on, 179
Quasiorthogonal STBC (QOSTBC), 157

R

Radiated electric field strength, 86
Radio-based CBTC projects, 8, 9*t*–12*t*

Radio channel with leaky waveguide, CBTC
modeling of, 84–91
measurement results, 84–85
path loss exponent, 85–88
small-scale fading, 88–91
overview, 82–83
Radio tests, 31–32
Rail and fastening parts
comparison of NDT methods, 57, 61*t*
infrastructure, 46–50
critical place on rail network, 49–50
fastening parts overview, 49
maintenance, 46
rail overview, 46–49
superstructure subsection, 46, 47*f*
inspection, 50–57
ACFM, 57
advantages and disadvantages of, 57,
58*t*–60*t*
eddy current, 56
liquid penetrant, 52
magnetic flux leakage, 55–56
manual and automated visual, 51–52
ultrasonic, 52–55, 53*f*
using EMATs, 57
Rail defects
broken rail, 48*f*
crack propagation, 47, 48*f*
cracks by manufacturing defects, 47
critical, 47
inappropriate handling and use, 49
loss of rail due to corrosion, 48*f*
noncritical, 47
rail wear and fatigue, 49
shelling, 48*f*
Rail-guided transport systems, 178
Rail system, traditional, 178
Railway signaling systems, 45
Random backoff scheme, 186
Random transmission errors, 182
packet drops due to, 186–187
Rayleigh channel, 128
Receiver sensitivity, 189
Reduced-state Bellman equation, 135
Regenerative period, 136
Reinforcement learning (RL) model, 215, 219*f*
algorithms, 220
in cognitive control approach, 225, 225*f*
objective of, 219
Reliability, availability, and maintenance tests,
36–37
Reward function, 131

Rigid fastenings, 49, 50*f*
Rolling contact fatigue (RCF), 49
Rolling stock characterization tests, 27–28

S

Safe braking distance (SBD), train's, 2
Second-order AIC, 88
Sectorial scanning, 54
Selective-repeat (SR)-ARQ, 128
Semi-Markov decision process (SMDP), 118
elements, 130–134
generalizes MDP, 130
optimal policy for, 134
policy, 235, 236*f*
handoff performance, 240
train travel trajectory under, 235,
237*f*, 239*f*
Sensor–controller delays, 183
Sensor-on-train system, 55
Service set identifier (SSID), 97–98, 159
Shadow effect, 188
Shadow mode tests, 36
Shark-fin antenna, 67–68, 70*f*
Short interframe space (SIFS), 155
application of, 160
Short view (Sv) scheme, 202
cost of train using, 208, 209*f*
performance of trains using, 204, 205*f*, 207*f*
Signal detection algorithm, 154–155
Signal passed at danger (SPAD), 44
Signal-to-noise ratio (SNR), 67, 97, 151, 221
amplitude of, 71
distribution of, 74–76
level thresholds of FSMC model, 72–74,
76*t*, 77*t*
probability distribution function of, 74
received, 69
Simple linear algorithms, 156
Site acceptance tests, 35–36
Small-scale area (SSA), 88
Small-scale fading model, 82
determination of, 88–91
empirical CDFs of, 88, 89*f*
SMDP-based CoMP cluster selection, 129–134
actions, 130–131
constraints, 134
decision epochs, 130
reward function, 131–132
solutions to, 134–137
online value iteration algorithm, 135–137
reduced-state Bellman equation, 135

states, 131
state transition probability, 132–134
SMDP-based optimization algorithm, 124–125, 129
Solid State Interlocking (SSI), 25, 39
Space–time block code (STBC), 151
  Alamouti codes, 158
  orthogonal, 157
  quasiorthogonal, 157
Spatial multiplexing gain, 224
State probability, 72
State space model, 124, 222
State transition probability, 72, 132–134
  matrix, 70–71, 223–224
Static PICO test, 28
Station adapter (SA), 96, 140
Stationary control policy, 134
Steady-state probability, 102
  for EMC, 109
  of MRGP, 110
Stochastic Petri net (SPN) model, 95
  generalized, 103–104
Stochastic process, 105
Sufficient information, 217
Supplier internal factory testing, CBTC, 26
Synchronized downlink transmission, 156, 156*f*
System models, 122–125
  communication channel model, 125
  train control model, 123–125, 123*f*
System optimization process, 139
System redundancy and recovery mechanisms, 7
System regression tests, 29
System tests, 29

T

Target equipment, tests on, 25
Testing/tests, CBTC, 17
  duration, 38
  environmental, 23–24
  factory, 24–27
  impacts on, 22, 39
  integration, 26, 32–33
  on-board integration, 27–28
  on-site, 30–37
  principles of
    in factory, 22
    reuse from previous projects, 21–22
    safety-related items, 22
Test track, 28–30
  equipment, 29
  location of, 29–30
  use of, 28–29

Track-based train control system, 178
Train
  control system objectives, 183
  dynamics
    model of, 221–222
    parameters, 231–232
  following model, 180, 181*f*
  LMAs of leading, 182
  state space equation, 231
  system to control group of, 181, 182*f*
  travel trajectories of, 235, 237*f*, 238*f*, 239*f*
Train control model, 123–125, 123*f*,
    221–223
Train control system
  evolution of, 2–5
    CBTC, 4–5, 4*f*
    profile-based, 3, 4*f*
  objective of, 2
  over wireless networks, 7
  performance improvement, 140–143
Train–ground communication systems, 126
  communication latency in, 119, 129
  derive availability of, 241
  failure rate, 240, 241*f*
  packet drops in, 186–193
    due to handover, 187–193
    due to random transmission errors,
      186–187
  QoS of, 199
  radio, 7
  related work, 179–180
  uncertainties in, 182
Train-mounted and walking stick probes,
    52, 53*f*
Train protection (TP) equipment, 45
Trains' control in CBTC systems, 180–185
  analytical formulation of, 184–185
  communication procedures with packet
    drops, 180–181, 181*f*
  and equivalent NCS, 181–184
  field test and simulation results, 201–208
    closed-loop control systems, 202–203
    impacted by packet drops, 203–208
    on packet drop rate, 201, 201*t*
    simulation parameters, 202, 202*t*
  with packet drops, 193–201
    currently used control scheme,
      193–194
    novel control schemes, 199–201
    on performances of, 198–199
    on stability of, 197–198
    states estimation under, 194–197

Train signaling system
  evolution of, 2–5
    distance-to-go technology, 3
    in-cab signals, 3
    using wayside signals, 2, 2f
  objective of, 2
Train's safe braking distance (SBD), 2
Train state space equation, 123
Train travel time error, 141, 143f
Train travel trajectory
  in CBTC system, 141, 142f
  under cognitive control policy, 235, 238f
  under greedy policy, 235, 237f, 239f
  under SMDP policy, 235, 237f, 239f
Transducer mounting for rail inspection, 55, 55f
Transition probability matrix, 108
Transmission interference period, 153
Transmission interruption period, 153–154
Transmission loss of leaky waveguide, 85–86
Transmitting antenna, 67–68
Trip error, 120, 121f
Tunnel
  Beijing Subway Changping Line, 68, 70f, 76
  to capture characteristics of, 69
  free space, 214
  section and deployment of antennas, 68, 69f

U

Ultrasonic rail inspection, 52–55
  conventional, 53
  geometrical limitations on, 52, 53f
  laser, 53–54
  long-range, 54–55
  phased array, 54
Unattended train operations (UTOs), 5
Unified modeling language (UML) software, 95
Uplink (UL) data, 155, 156f
Uplink delay, 229–230
Uplink transmissions, 183–184
Urban rail transit systems, 82
  key subsystem of, 214
  modeling channels of, 66
  WLANs, 178

V

V-BLAST algorithm, 155, 157
Vehicle onboard components
  (VOBCs), 183
Velocity deviation, 185
Velocity tracking error, 124, 131–133
Visual rail inspection, 51–52

W

Wayside ATO system, 6
Wayside ATP system, 6
Wayside network tests, 30–31
Wireless channel model, 162–163
  parameters, 233–234, 235t
Wireless local area network (WLAN),
  66, 94
  APs, 66
  based DCS, 94
    availability analysis, 99–102
    basic configuration of, 96, 96f
    behavior with DSPNs,
      103–110
    first proposed, 97, 98f
    overview of CBTC and, 96–97
    with redundancy, 97–99
    second proposed, 98, 99f
    unavailability of, 111, 112f
    wireless portions of, 96
  channel model in MIMO-enabled, 223–225
  communication latency in, 161–162
  handoff procedure, 228, 228f
  handoff timing diagram, 152, 152f
  IEEE 802.11g, 226
  related works, 95

Y

Yagi antenna, 67–68, 70f, 234f

Z

Zone controller (ZC), 21, 25, 96, 119, 217–218